Selected Titles in This Series

664 **Xia Chen,** Limit theorems for functionals of ergodic Markov chains with general state space, 1999

663 **Ola Bratteli and Palle E. T. Jorgensen,** Iterated function systems and permutation representation of the Cuntz algebra, 1999

662 **B. H. Bowditch,** Treelike structures arising from continua and convergence groups, 1999

661 **J. P. C. Greenlees,** Rational S^1-equivariant stable homotopy theory, 1999

660 **Dale E. Alspach,** Tensor products and independent sums of $\mathcal{L}_p$-spaces, $1 < p < \infty$, 1999

659 **R. D. Nussbaum and S. M. Verduyn Lunel,** Generalizations of the Perron-Frobenius theorem for nonlinear maps, 1999

658 **Hasna Riahi,** Study of the critical points at infinity arising from the failure of the Palais-Smale condition for n-body type problems, 1999

657 **Richard F. Bass and Krzysztof Burdzy,** Cutting Brownian paths, 1999

656 **W. G. Bade, H. G. Dales, and Z. A. Lykova,** Algebraic and strong splittings of extensions of Banach algebras, 1999

655 **Yuval Z. Flicker,** Matching of orbital integrals on $GL(4)$ and $GSp(2)$, 1999

654 **Wancheng Sheng and Tong Zhang,** The Riemann problem for the transportation equations in gas dynamics, 1999

653 **L. C. Evans and W. Gangbo,** Differential equations methods for the Monge-Kantorovich mass transfer problem, 1999

652 **Arne Meurman and Mirko Primc,** Annihilating fields of standard modules of $\mathfrak{sl}(2,\mathbb{C})^\sim$ and combinatorial identities, 1999

651 **Lindsay N. Childs, Cornelius Greither, David J. Moss, Jim Sauerberg, and Karl Zimmermann,** Hopf algebras, polynomial formal groups, and Raynaud orders, 1998

650 **Ian M. Musson and Michel Van den Bergh,** Invariants under Tori of rings of differential operators and related topics, 1998

649 **Bernd Stellmacher and Franz Georg Timmesfeld,** Rank 3 amalgams, 1998

648 **Raúl E. Curto and Lawrence A. Fialkow,** Flat extensions of positive moment matrices: Recursively generated relations, 1998

647 **Wenxian Shen and Yingfei Yi,** Almost automorphic and almost periodic dynamics in skew-product semiflows, 1998

646 **Russell Johnson and Mahesh Nerurkar,** Controllability, stabilization, and the regulator problem for random differential systems, 1998

645 **Peter W. Bates, Kening Lu, and Chongchun Zeng,** Existence and persistence of invariant manifolds for semiflows in Banach space, 1998

644 **Michael David Weiner,** Bosonic construction of vertex operator para-algebras from symplectic affine Kac-Moody algebras, 1998

643 **Józef Dodziuk and Jay Jorgenson,** Spectral asymptotics on degenerating hyperbolic 3-manifolds, 1998

642 **Chu Wenchang,** Basic almost-poised hypergeometric series, 1998

641 **W. Bulla, F. Gesztesy, H. Holden, and G. Teschl,** Algebro-geometric quasi-periodic finite-gap solutions of the Toda and Kac-van Moerbeke hierarchies, 1998

640 **Xingde Dai and David R. Larson,** Wandering vectors for unitary systems and orthogonal wavelets, 1998

639 **Joan C. Artés, Robert E. Kooij, and Jaume Llibre,** Structurally stable quadratic vector fields, 1998

638 **Gunnar Fløystad,** Higher initial ideals of homogeneous ideals, 1998

637 **Thomáš Gedeon,** Cyclic feedback systems, 1998

636 **Ching-Chau Yu,** Nonlinear eigenvalues and analytic-hypoellipticity, 1998

(*Continued in the back of this publication*)

Limit Theorems for Functionals of Ergodic Markov Chains with General State Space

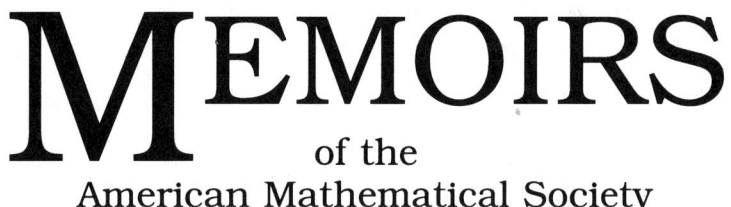
of the
American Mathematical Society

Number 664

Limit Theorems for Functionals of Ergodic Markov Chains with General State Space

Xia Chen

May 1999 • Volume 139 • Number 664 (third of 5 numbers) • ISSN 0065-9266

American Mathematical Society
Providence, Rhode Island

1991 *Mathematics Subject Classification.*
Primary 60B12, 60F05, 60F10, 60F15, 60J10.

Library of Congress Cataloging-in-Publication Data
Chen, Xia, 1956–
    Limit theorems for functionals of ergodic Markov chains with general state space / Xia Chen.
    p. cm. — (Memoirs of the American Mathematical Society, ISSN 0065-9266 ; no. 664)
    "May 1999, volume 139, number 664 (third of 5 numbers)."
    Includes bibliographical references.
    ISBN 0-8218-1060-X (alk. paper)
    1. Central limit theorem.   2. Markov processes.   3. Large deviations.   I. Title.   II. Series.
QA3.A57 no. 664
[QA273.67]
510s—dc21
[519.2]   99-19209
                                                                                                                                                                                        CIP

Memoirs of the American Mathematical Society
    This journal is devoted entirely to research in pure and applied mathematics.

    Subscription information. The 1999 subscription begins with volume 137 and consists of six mailings, each containing one or more numbers. Subscription prices for 1999 are $448 list, $358 institutional member. A late charge of 10% of the subscription price will be imposed on orders received from nonmembers after January 1 of the subscription year. Subscribers outside the United States and India must pay a postage surcharge of $30; subscribers in India must pay a postage surcharge of $43. Expedited delivery to destinations in North America $35; elsewhere $130. Each number may be ordered separately; *please specify number* when ordering an individual number. For prices and titles of recently released numbers, see the New Publications sections of the *Notices of the American Mathematical Society*.

    Back number information. For back issues see the *AMS Catalog of Publications*.

    Subscriptions and orders should be addressed to the American Mathematical Society, P. O. Box 5904, Boston, MA 02206-5904. *All orders must be accompanied by payment.* Other correspondence should be addressed to Box 6248, Providence, RI 02940-6248.

    Copying and reprinting. Individual readers of this publication, and nonprofit libraries acting for them, are permitted to make fair use of the material, such as to copy a chapter for use in teaching or research. Permission is granted to quote brief passages from this publication in reviews, provided the customary acknowledgment of the source is given.

    Republication, systematic copying, or multiple reproduction of any material in this publication (including abstracts) is permitted only under license from the American Mathematical Society. Requests for such permission should be addressed to the Assistant to the Publisher, American Mathematical Society, P. O. Box 6248, Providence, Rhode Island 02940-6248. Requests can also be made by e-mail to reprint-permission@ams.org.

    Memoirs of the American Mathematical Society is published bimonthly (each volume consisting usually of more than one number) by the American Mathematical Society at 201 Charles Street, Providence, RI 02904-2294. Periodicals postage paid at Providence, RI. Postmaster: Send address changes to Memoirs, American Mathematical Society, P. O. Box 6248, Providence, RI 02940-6248.

    © 1999 by the American Mathematical Society. All rights reserved.
This publication is indexed in *Science Citation Index*®, *SciSearch*®, *Research Alert*®, *CompuMath Citation Index*®, *Current Contents*®/*Physical, Chemical & Earth Sciences*.
Printed in the United States of America.

    ∞ The paper used in this book is acid-free and falls within the guidelines
established to ensure permanence and durability.
Visit the AMS home page at URL: http://www.ams.org/

10 9 8 7 6 5 4 3 2 1   04 03 02 01 00 99

CONTENTS

Introduction — xi

Chapter I. Split Chain and Regeneration — 1

    I-1 Motivation and notation — 1

    I-2 Split chain — 8

    I-3 Maximal and minimal inequalities — 19

    I-4 Maximal integrability on small sets — 24

Chapter II. The Central Limit Theorem — 27

    II-1 Introduction — 27

    II-2 Validity of CLT — 29

    II-3 Identification of the limiting variance in the CLT — 38

    II-4 The CLT and the ergodicity conditions — 58

Chapter III. The Law of the Iterated Logarithm — 67

    III-1 Introduction — 67

    III-2 The LIL and CLT — 69

    III-3 Some remarks on the limit set — 83

    III-4 The LIL in vector setting — 91

    III-5 Some consequences — 106

    III-6 Supplement — 113

Chapter IV. The moderate Deviation Principle — 119

    IV-1 Introduction — 119

    IV-2 Upper bounds — 123

    IV-3 Lower bounds — for chains with a small set of order 1 — 134

IV-4 Lower bounds — when ξ is bounded	147
IV-5 Lower bounds — when $\xi \in CLT$	162
Appendix	172
References	200

ABSTRACT

Let $\{X_n\}_{n\geq 0}$ be an ergodic Markov chain with general state space E and let ξ be a measurable map from E to $\mathbf{R}$ or to some separable Banach space B and set

$$S_n = \sum_{k=0}^{n-1} \xi(X_k) \quad n = 1, 2, \cdots.$$

The central limit theorem, the law of the iterated logarithm and the moderate deviation principle associated with $\{S_n\}_{n\geq 1}$ are established in a reasonable and natural way. Our arguments mainly depend on the split chain and regeneration methods which are systematically developed in this paper. All conditions appear to be the best possible (some even necessary) for our theorems and what is important, they are given in terms of the original chain.

AMS 1991 subject classifications. 60B12, 60F05, 60F10, 60F15, 60J10.

Key words and phrases. ergodic Markov chain, invariant distribution, atom, split chain, central limit theorem, law of the iterated logarithm, moderate deviation.

To Lin and Amy

INTRODUCTION

One of the most important problems in probability theory is the investigation of the limit theorems for appropriately normalized sums of random variables. The case of independent random variables is fairly well understood, but less is known about dependent random variables such as Markov chains, one of the most important generalizations of the independent case. The purpose of this work is to study several basic limit theorems for functionals of Markov chains.

With the publication of the book of Chung (1967), a method known as regeneration made it technically possible to establish the central limit theorem and the law of the iterated logarithm for functionals of stationary Markov chains with a countable state space. The main idea of this method is to analyze the Markov chain by dividing it into independent and identically distributed (i.i.d.) random blocks between visits to a fixed state (or atom). Independently working on the same problem in 1978, Athreya-Ney ([8]) and Nummelin ([38]) introduced, in the general (possibly non-atomic) state space case, the technique called split chain method, a way to create an atom or a "pseudoatom" which allows for the regeneration. This method was refined later in [3], [32], [39]. For a systematic study of the splitting technique and regeneration phenomena, the reader is refered to two excellent books written by Nummelin ([39]) and Meyn-Tweedie ([33]). For the literature on the applications of the regeneration and split chain scheme, see [3], [4], [5], [8], [17], [18], [19], [33], [35], [36], [37], [38], [39], [43].

Received by the editor November 18, 1996, and in revised form December 28, 1997.

In this paper, some limit theorems for real or vector valued functionals of ergodic Markov chains with general state space are investigated. These are the central limit theorem (Chapter II), the law of the iterated logarithm of ergodic Markov chains with general state space are investigated. These are the central limit theorem (Chapter II), the law of the iterated logarithm (Chapter III) and the moderate deviation principle (Chapter IV). The split chain and regeneration scheme, which is essential for the establishment of these theorems, is extended, modified and clarified (Chapter I) in an analytic way. To describe what this paper tries to achieve, I would like to mention a few facts.

It is the aim of this paper to state all theorems in terms of the original chain. This is accomplished with a few exceptions (Theorem II-2.2 and Theorem II-2.3), where the split chain structure is preserved in the statement of results for some technical reasons. Indeed, this paper tries to solve a problem quite common in the applications of the split chain and regeneration scheme rising from the use of the split chain technique, with which the regeneration is applied to some artificially constructed chain rather than the original one. That is, the theorems need to be formulated based on the data directly from the original chain rather than from any artificial construction.

This paper tries to identify various limiting parameters in the theorems under some natural, mild conditions. In Theorem II-3.1, the limiting variance in the central limit theorem is identified in a natural way. Theorem II-3.1 appears to be a sharp formulation of the central limit theorem. More importantly, it changes the traditional way of formulating limit theorems for Markov chains, where the manner in which a theorem is stated — in particular, the form of the limiting variance in the central limit theorem — usually depends on the way this theorem is proved. The identification problems for the other limit theorems are thus solved as corollories of Theorem II-3.1 due to their connection (as shown in the paper) with the central limit theorem.

Sharpness is another goal this paper tries to pursue. As mentioned before, the central limit theorem is established under sharp conditions. Much like the i.i.d. case, necessary and sufficient conditions are given for the law of the iterated logarithm (Theorem III-4.3). The comparison between the central limit theorem and the law of the iterated logarithm (Section III-2) shows that the law of the iterated logarithm implies the central limit theorem but the converse is not true. Theorem III-4.3 (or its corollory) also shows exactly how much additional information is needed for the law of the iterated logarithm if one has the central limit theorem. The conditions for the upper bounds of the moderate deviations (Theorem IV-2.1) appear to be minimal. To establish the lower bound under minimal assumptions on the Markov kernel and the functional appears to be a difficult problem. We have succeeded in proving three non-trivial lower bound results under different, but fairly weak conditions (Theorem IV-3.1, Theorem IV-4.1 and Theorem IV-5.1). Despite the difficulty to unify them, each is a natural generalization of the independent case. Besides, this is the first time that this type of result has been established without further ergodicity assumptions.

In addition to the traditional way of using the split chain and regeneration method, this paper presents some useful ideas and tools. Some of them have appeared in other fields of probabilty. The maximal inequality is introduced to control the trajectory of the regeneration sequence. Its dual, the minimal inequality, is applied to the lower bound problem of moderate deviations. We would like to highlight a dichotomy result (Theorem I-4.1), which plays a fundamental role in describing the strong limit theorems for ergodic Markov chains. The surprising fact revealed by Theorem I-4.1 may have other applications in Markov chain theory. The randomization method (Lemma II-3.3), a way to preserve certain limiting behavior through the resolvent approximation, reduces some problems to a relatively simpler situation. The idea of directly using the minorization of the Markov kernel, which first appeared in [22], takes an important part in this work. Through the regeneration procedure, some powerful tools used in the independent context, such as Wald's equation, Lévy decomposition and isoperimetric inequality, are also used for certain problems in this paper.

I want to express my sincere thanks to my advisor Professor Alejandro de Acosta. His excellent teaching of the course on Markov chains and his unpublished notes on the regeneration method and the central limit theorem for Markov chains greatly inspired my interest in this subject. His analytic way ([3]) of handling the split procedure has a strong influence on the present paper. His careful reading of this manucript and his many advices led to a substantial improvement of this paper. I would also like to thank Professor J. Kuelbs, for his careful reading of the original manuscript and for his many good suggestions. Finally, I am indebted to the referee of this paper for some very valuable comments and suggestions.

Chapter I. Split Chain and Regeneration

I-1. Motivation and notation

The basic technique in the study of limit theorems for a Markov chain with an atom is the regeneration method, which consists in dividing the chain into interblocks between its returns to the atom (we refer to the book of Chung (1967) for the details). These interblocks are independent and identically distributed (i.i.d.). When the chain has no atom, one may attempt to approximate this method by replacing the atom by a suitably chosen "small set". However, the interblocks between the visits to the small set are not necessarily independent. Independently initiated by Nummelin (1978, 1984) and Athreya-Ney (1978) and later refined by de Acosta (1988) and Meyn-Tweedie (1993), a splitting technique has been developed for some Markov chain with general (possibly nonatomic) state space, which embeds the chain into its split chain to create an "artificial" atom. For obvious reasons, the splitting technique, as well as the regeneration method, becomes a powerful tool in the study of limit theorems for Markov chains with general state space.

We devote this chapter to study the split and regeneration phenomena, which are crucial for the later development of the paper. In Section I-2, we construct the split chain based on the same idea as Nummelin (1984) but in a more analytic way. Since the proof of our limit theorems largely relies on transferring various objects and concepts between the original and the split chains, effective computability of the splitting technique is required for our goals. In Section I-3, we establish the maximal and minimal inequalities for the Markov chains before and after splitting. We will see in the later development how these inequalties greatly enhance the power of the regeneration method. In Section I-4, we discuss the "maximal" integrability on small sets, which turns out to be fundamental in the study of strong limit theorems and has some interesting applications to the classical Markov chain theory as well. The main achievement in this section is a dichotomy theorem about this integrability.

We skip the proofs of some results in this chapter in the hope that the reader may get into our main theorems given in the later chapters before interrupted by too many technical details. Instead, these proofs will be given in an appendix at the end of this paper.

The following notations and concepts, which are mainly adopted from Nummelin (1984) and Meyn-Tweedie (1993), will be used throughout the paper. In this work, $(E, \mathcal{E})$ is a measurable space with the σ-algebra $\mathcal{E}$ countably generated. Given a bounded signed measure μ on $(E, \mathcal{E})$, $||\mu||_V$ denotes its total variation, i.e.

$$(1.1) \qquad ||\mu||_V = \sup_{||h||_\infty \leq 1} \left| \int h(x) \mu(dx) \right|.$$

Consider a Markov chain $\{X_n\}_{n \geq 0}$ with the state space $(E, \mathcal{E})$, the transition probability $P(x, A)$ and the n-step transition probability $P^n(x, A)$ for each $n \geq 1$.

Given an initial distribution, namely a probability measure μ on $(E, \mathcal{E})$ which is the distribution of X_0, the probability distribution of the chain $\{X_n\}_{n \geq 0}$, which is denoted by P_μ, is uniquely determined by $P(x, A)$ and μ such that

$$P_\mu \{X_0 \in A_0, X_1 \in A_1, \cdots, X_n \in A_n\}$$
$$= \int_{A_0} \mu(dx_0) \int_{A_1} P(x_0, dx_1) \cdots \int_{A_{n-1}} P(x_{n-1}, A_n)$$

holds for all $A_0, A_1, \cdots, A_n \in \mathcal{E}$ and all $n \geq 0$. The corresponding expectation will be denoted by E_μ. When $\mu = \delta_x$ is the Dirac measure at x, P_μ and

E_μ will be denoted by P_x and E_x, respectively.

Although no specification is made in our main theorems for the probability space $(\Omega, \mathcal{F}, P_\mu)$ on which the chain $\{X_n\}_{n\geq 0}$ is defined, it is often convenient to think of $(\Omega, \mathcal{F})$ as the product measurable space $(\prod_{n=0}^\infty E_n, \otimes_{n=0}^\infty \mathcal{E}_n)$, where $(E_n, \mathcal{E}_n)$ is a copy of $(E, \mathcal{E})$, and $\{X_n\}_{n\geq 0}$ as the canonical projections on Ω. Hence we do this without further comment whenever it is necessary. Under such identification, the **shift operator** (w.r.t. $\{X_n\}_{n\geq 0}$) $\theta\colon \Omega \longrightarrow \Omega$ defined by

$$\theta(\{x_0, x_1, \cdots, x_n, \cdots\}) = \{x_1, x_2, \cdots, x_{n+1}, \cdots\} \quad \text{where} \quad \omega = \{x_n\}_{n\geq 0} \in \Omega$$

satisfies

$$\{X_0, X_1, \cdots, X_n, \cdots\} \circ \theta = \{X_1, X_2, \cdots, X_{n+1}, \cdots\}.$$

We write θ^n for the nth power of θ: $\theta^n = \overbrace{\theta \circ \theta \circ \cdots \circ \theta}^{n}$.

The chain $\{X_n\}_{n\geq 0}$ (or its transition P) is called **irreducible** if there exists a σ-finite measure ψ on $(E, \mathcal{E})$ which we call an **irreducibility measure** for $\{X_n\}_{n\geq 0}$, such that for every $A \in \mathcal{E}$,

$$\psi(A) > 0 \Rightarrow \sum_{n=1}^\infty P^n(x, A) > 0 \quad \text{for all} \quad x \in E.$$

It is assumed throughout this study that all Markov chains we deal with are irreducible. According to Proposition 2.4 of Nummelin (1984), there exists a measure ψ on $(E, \mathcal{E})$ which we call **maximal irreducibility measure** for $\{X_n\}_{n\geq 0}$, such that ψ is an irreducibility measure and that all other irreducibility measures are absolutely continuous w.r.t. ψ. Write

$$\mathcal{E}^+ = \{A \in \mathcal{E};\ \psi(A) > 0\},$$

where ψ is a maximal irreducibility measure. Note that, by definition, a maximal irreducibility measure is unique up to the equivalence of measures. Therefore $\mathcal{E}^+$ is uniquely defined.

The chain $\{X_n\}_{n\geq 0}$ is called **recurrent** if

$$\sum_{n=1}^{\infty} P^n(x, A) = +\infty$$

for all $x \in E$ and all $A \in \mathcal{E}^+$; $\{X_n\}_{n\geq 0}$ is called **Harris recurrent** if

$$P_x\{X_n \in A \ i.o.\} = 1$$

for all $x \in E$ and all $A \in \mathcal{E}^+$. By the Borel-Cantelli lemma, recurrence is implied by Harris recurrence. However, the converse may not be true in the general state space context (see p.204 of Meyn-Tweedie (1993) for an example).

Given $A \in \mathcal{E}^+$, τ_A denotes the **first return time** to A, i.e.

(1.2) $$\tau_A = \inf\{n \geq 1; X_n \in A\}.$$

Then, τ_A is a stopping time w.r.t. $\{X_n\}_{n\geq 0}$ and Harris recurrence implies that $\tau_A < +\infty$ a.s. w.r.t. P_μ for every initial distribution μ.

A σ-finite measure π (finite or not) is **invariant** if $\pi = \pi P$, that is,

$$\pi(A) = \int \pi(dx) P(x, A) \qquad A \in \mathcal{E}.$$

According to Theorem 10.4.4 of Meyn-Tweedie (1993), the recurrence of the chain implies the existence and uniqueness (up to constant multiples) of an invariant measure. In this case $\{X_n\}_{n\geq 0}$ is called **positive** (**null**, resp.) if the invariant measure is finite (infinite, resp.). The symbol π will be used exclusively to denote an invariant measure and when $\{X_n\}_{n\geq 0}$ is positive,

π always denotes the one with $\pi(E) = 1$, in which case we use the name **invariant distribution** to emphasize this choice.

The concept of atom plays a crucial role in this paper. A set $\alpha \in \mathcal{E}^+$ is called an **atom** of $\{X_n\}_{n \geq 0}$ (or its transition P) if

(1.3) $$P(x, \cdot) = P(y, \cdot) \qquad \text{for all } x, y \in \alpha.$$

Noticing that (1.3) implies that $P_x = P_y$ on the σ-algebra generated by $\{X_n\}_{n \geq 1}$, we denote the common value by P_α. Notations like $P(\alpha, \cdot)$ and E_α are also used in the obvious way.

Atoms do not always exist in a general state space. So we use small sets instead. A set $C \in \mathcal{E}^+$ is called **small** if the following minorization

(1.4) $$P^m \geq b I_C \otimes \nu$$

holds for some $m \geq 1$, some $b > 0$ and some probability measure ν on $(E, \mathcal{E})$. The **order** of a small set C is the smallest m satisfying (1.4) with some $b > 0$ and some probability measure ν. A set $C \in \mathcal{E}$ is called **sub-π small** if $0 < \pi(C) < +\infty$ and C satisfies (1.4) with ν being specified as π_C:

(1.5) $$\pi_C(A) \equiv \frac{1}{\pi(C)} \pi(C \cap A) \qquad \forall A \in \mathcal{E}.$$

Clearly, an atom is a small set of order 1. Since (Theorem 10.4.9 of Meyn-Tweedie (1993)) π is a maximal irreducibility measure when $\{X_n\}_{n \geq 0}$ is recurrent, a sub-π small set is a small set. The theorems in Section 5.2 of Meyn-Tweedie (1993) show that small sets not only exist, but exist in abundance. In particular, Theorem 5.2.1 (with $\psi = \pi$) implies the existence of sub-π small sets.

Now, fix a small set C together with a probability measure ν on $(E, \mathcal{E})$ such that (C, ν) satisfies (1.4) for some $m \geq 1$ and that $\nu(C) > 0$ (such (C, ν)

exists by Theorem 5.2.2 in Meyn-Tweedie (1993)). Let $d \geq 1$ be the greatest common divisor of the set

$$\{m \geq 1; \text{ there exists } b > 0 \text{ such that } (1.4) \text{ holds}\}.$$

By Theorem 2.2 of Nummelin (1984), d does not depend on the particular choice of (C, ν). We call d the **period** of the chain $\{X_n\}_{n \geq 0}$ (for a probabilistic interpretation of the period of Markov chains, see Theorem 2.2 of Nummelin (1984)). The chain $\{X_n\}_{n \geq 0}$ is called **aperiodic** if $d = 1$.

The chain $\{X_n\}_{n \geq 0}$ (or its transition P) is called **ergodic** if it is positive, Harris recurrent and aperiodic. According to Proposition 6.3 of Nummelin (1984), a Markov chain $\{X_n\}_{n \geq 0}$ is ergodic if and only if $\{X_n\}_{n \geq 0}$ is positive and

$$(1.6) \qquad \lim_{n \to \infty} ||P^n(x, \cdot) - \pi||_V = 0 \quad \forall x \in E.$$

This reveals that the ergodicity of $\{X_n\}_{n \geq 0}$ is nothing but its asymptotic stationarity. In addition to this "basic" ergodicity, ergodicities of different levels are discussed in Nummelin (1984) and Meyn-Tweedie (1993). The chain $\{X_n\}_{n \geq 0}$ is called **ergodic of degree 2** (Proposition 5.16 and Section 6.4 of Nummelin (1984)) if it is ergodic and

$$(1.7) \qquad E_\pi \tau_A < +\infty \quad \forall A \in \mathcal{E}^+;$$

$\{X_n\}_{n \geq 0}$ is called **geometrically ergodic** (Theorem 6.14 of Nummelin (1984)) if it is ergodic and there exists $r_o > 1$ such that

$$(1.8) \qquad \sum_{n=1}^{\infty} r_o^n \int ||P^n(x, \cdot) - \pi||_V \pi(dx) < +\infty;$$

$\{X_n\}_{n \geq 0}$ is called **uniformly ergodic** (p. 383 of Meyn-Tweedie (1993)) if

$$(1.9) \qquad \lim_{n \to \infty} \sup_{x \in E} ||P^n(x, \cdot) - \pi||_V = 0.$$

According to Theorem 16.02 of Meyn-Tweedie (1993), uniform ergodicity means that there exists $r > 1$ such that

$$\sup_{x \in E} ||P^n(x, \cdot) - \pi||_V \leq r^{-n}$$

holds for sufficiently large n, which implies the geometric ergodicity. In Chapter II we will prove that the chain $\{X_n\}_{n \geq 0}$ is ergodic of degree 2 if and only if it is ergodic and

$$\sum_{n=1}^{\infty} \int ||P^n(x, \cdot) - \pi||_V \pi(dx) < +\infty.$$

Therefore, we have the following relations among these ergodic properties:

uniform ergodicity $\implies$ geometric ergodicity
$\implies$ ergodicity of degree 2 $\implies$ ergodicity.

The splitting technique, a procedure to create atoms in general state space context, can make an atom only for the chain with a small set of order 1. It is desirable to know whether or not that the small set of order 1 always exists under some resonable conditions. Unfortunately, the answer is negative even forsome "nice" Markov chains. Here is an example.

Example 1.1. Let $(M, \mathcal{M})$ be a measurable space with $\mathcal{M}$ countably generated and let μ be a probability measure on $(M, \mathcal{M})$ such that

(1.10) $$\mu(\{x\}) = 0 \quad \forall x \in M.$$

Let $(E, \mathcal{E}) \equiv (M \times M, \mathcal{M} \otimes \mathcal{M})$ be the product measurable space and define the Markov transition P on E as follows:

$$P((x, y), \cdot) = (\delta_y \otimes \mu)(\cdot) \quad (x, y) \in M \times M.$$

One can verify that $\pi \equiv \mu \otimes \mu$ is the invariant distribution of P and $P^2 = \pi$. Therefore P is uniformly ergodic and E is a small set of order 2. On the other hand, (1.4) with m=1 implies that the set C must be of the form:

$$C = A \times \{y_o\}.$$

From (1.10) we have $\pi(C) = 0$. Note the fact that π is a maximal irreducibility measure. Hence $C \notin \mathcal{E}^+$. ∎

In spite of this, by the splitting technique we can create a "pseudo-atom" in the general case ($m \geq 1$), which has some properties similar to those of an atom. Additionally, a chain (possibly without a small set of order 1) can be approximated very well in some situations by chains with the small sets of order 1. However, the gap still exists and the resulting difficulties sometimes are essential (see the establishment of the lower bounds for moderate deviations in Chapter IV for example). We leave this for future study.

I-2. Split chain

In this section, we try to embed $\{X_n\}_{n \geq 0}$ into a larger probability space on which a $\{0,1\}$-valued random sequence $\{Y_k\}_{k \geq 0}$ are also defined. Furthermore, the embedding will be carried out in such a way that the following two random sequences (where $m \geq 1$ is given in (1.4))

$$\{(X_{(n-1)m+1}, \cdots, X_{nm}, Y_n)\}_{n \geq 1} \quad \text{and} \quad \{(X_{nm}, Y_n)\}_{n \geq 0}$$

become Markov chains which inherit some important properties possessed by the original chain $\{X_n\}_{n \geq 0}$ (see Theorem 2.1 and Theorem 2.5). Most importantly, these resulting Markov chains have their atom or, "psuedo-atom" at least, with which the chain $\{X_n\}_{n\geq 0}$ is divided into independent or "1-dependent" interblocks by a regeneration argument (see Theorem 2.2, Corollary 2.3 and 2.4).

Our construction of the split chain is based on an observation of de Acosta (1988). Let $\{X_n\}_{n\geq 0}$ be an irreducible Markov chain with state space $(E, \mathcal{E})$ and transition $P(x, A)$ and assume (1.4) for fixed small set C, $m \geq 1$, $b > 0$ and probability measure ν. Without loss of generality we assume that $b < 1$. Define the Markov kernel Q on $(E, \mathcal{E})$ as follows:

$$Q(x, A) = (1 - bI_C(x))^{-1}(P^m(x, A) - bI_C(x)\nu(A)) \quad x \in E, \quad A \in \mathcal{E}.$$

Then,

$$(2.1) \quad P^m(x, A) = (1 - bI_C(x))Q(x, A) + bI_C(x)\nu(A) \quad x \in E, \quad A \in \mathcal{E}.$$

Define the transition functions P_0 and P_1: $E \times \mathcal{E}^m \longrightarrow [0, 1]$ as follows:

$$P_0(x, \Delta) = \int_\Delta \mathcal{L}_{P_x}(X_1, \cdots, X_m)(d(x_1, \cdots, x_m)) \frac{dQ(x, \cdot)}{dP^m(x, \cdot)}(x_m),$$

where $x \in E$ and $\Delta \in \mathcal{E}^m$.

$$P_1(x, \Delta) = \begin{cases} \int_\Delta \mathcal{L}_{P_x}(X_1, \cdots, X_m)(d(x_1, \cdots, x_m)) \frac{d\nu}{dP^m(x, \cdot)}(x_m) & x \in C \\ \mathcal{L}_{P_\nu}(X_1, \cdots, X_m)(\Delta) & x \notin C \end{cases}$$

where $\Delta \in \mathcal{E}^m$ and we use $R(x) \equiv (d\mu_1/d\mu_2)(x)$ to denote the Radon-Nikodym derivative for given measures μ_1 and μ_2 on $(E, \mathcal{E})$ with $\mu_1 \ll \mu_2$. From (2.1),

$$(2.2) \quad \mathcal{L}_{P_x}(X_1, \cdots, X_m)(\cdot) = (1 - bI_C(x))P_0(x, \cdot) + bI_C(x)P_1(x, \cdot) \quad x \in E$$

(From (2.2), one can see that for $x \notin C$, the definition of $P_1(x, \Delta)$ has no contribution to the distribution $\mathcal{L}_{P_x}(X_1, \cdots, X_m)$, which we have chosen to be $\mathcal{L}_{P_\nu}(X_1, \cdots, X_m)$ simply to complete the definition).

Let $I = \{0, 1\}$ and let $\mathcal{I}$ be the trivial σ-algebra on I. Given a (probability) measure μ on $(E, \mathcal{E})$, define the (probability) measure μ^* on $(E \times I, \mathcal{E} \otimes \mathcal{I})$ as follows:

$$(2.3) \qquad \mu^* = \mu((1 - bI_C)I_{(\cdot)}) \otimes \delta_0 + \mu(bI_C I_{(\cdot)}) \otimes \delta_1,$$

where $\mu(hI_{(\cdot)})$ denotes, for given non-negative measurable function h on $(E, \mathcal{E})$, the measure

$$\mu(hI_{(\cdot)})(A) = \int_A h(x)\mu(dx) \qquad A \in \mathcal{E}.$$

Define the Markov kernel $\tilde{P}$: $(E \times I) \times (\mathcal{E}^m \otimes \mathcal{I}) \longrightarrow [0, 1]$ as follows:

$$(2.4) \qquad \begin{aligned} &\tilde{P}((x, y), \tilde{\Delta}) \\ &= I_0(y)\Big\{\Big(\int_{(\cdot)} P_0(x, d(x_1, \cdots, x_m))(1 - bI_C(x_m))\Big) \otimes \delta_0 \\ &\quad + \Big(\int_{(\cdot)} P_0(x, d(x_1, \cdots, x_m))bI_C(x_m)\Big) \otimes \delta_1\Big\}(\tilde{\Delta}) \\ &\quad + I_1(y)\Big\{\Big(\int_{(\cdot)} P_1(x, d(x_1, \cdots, x_m))(1 - bI_C(x_m))\Big) \otimes \delta_0 \\ &\quad + \Big(\int_{(\cdot)} P_1(x, d(x_1, \cdots, x_m))bI_C(x_m)\Big) \otimes \delta_1\Big\}(\tilde{\Delta}), \end{aligned}$$

where $(x, y) \in E \times I$ and $\tilde{\Delta} \in \mathcal{E}^m \otimes \mathcal{I}$.

Let μ be the initial distribution of $\{X_n\}_{n \geq 0}$. To justify our construction in the case $m > 1$ (we skip this step if $m = 1$), consider the enlarged space $\Omega' = \prod_{n=-(m-1)}^{\infty} E_n$, where E_n is a copy of E. Let $X_{-(m-1)}, \cdots, X_{-1}$ be the first $m - 1$ canonical projections on Ω' and define the probability distribution $P'_{\mu'}$ on Ω' by $P'_{\mu'} = \mathcal{L}_{P_\mu}(X_1, \cdots, X_{m-1}) \otimes P_\mu$. Under the law

$P'_{\mu'}$, $\{(X_{(k-1)m+1}, \cdots, X_{km})\}_{k \geq 0}$ becomes a Markov chain with the state space $(E^m, \mathcal{E}^m)$, the initial distribution μ' defined by

$$\mu' = \mathcal{L}_{P_\mu}(X_1, \cdots, X_{m-1}) \otimes \mu$$

and the transition probability R on $(E^m, \mathcal{E}^m)$ given by

(2.5) $$R\big((x_{-(m-1)}, \cdots, x_0), \Delta\big) = \mathcal{L}_{P_{x_0}}(X_1, \cdots, X_m)(\Delta),$$

where $(x_{-(m-1)}, \cdots, x_0) \in E^m$ and $\Delta \in \mathcal{E}^m$. That is,

$$P'_{\mu'}\{(X_{km+1}, \cdots, X_{(k+1)m}) \in \Delta | \mathcal{A}_k\} = P_{X_{km}}\{(X_1, \cdots, X_m) \in \Delta\} \quad a.s.$$

for each $k \geq 0$, where $\Delta \in \mathcal{E}^m$ and $\{\mathcal{A}_k\}_{k \geq 0}$ is the canonical filtration of this Markov chain.

On the other hand, write

$$Z_n = (\tilde{X}_{(n-1)m+1}, \cdots, \tilde{X}_{nm}, Y_n) \quad n = 0, 1, \cdots$$

to denote the canonical projections on $\prod_{n=0}^{\infty}(E^m \times I)$ and let $\tilde{P}_{\tilde{\mu}}$ be the Markovian distribution on $\prod_{n=0}^{\infty}(E^m \times I)$ determined by the transition probability $\tilde{R}$ on $(E^m \times I, \mathcal{E}^m \otimes \mathcal{I})$ given by

$$\tilde{R}(z, \tilde{\Delta}) = \tilde{P}(\Phi(z), \tilde{\Delta}), \quad z \in E^m \times I, \ \tilde{\Delta} \in \mathcal{E}^m \otimes \mathcal{I},$$

where $\Phi: E^m \times I \longrightarrow E \times I$ is given by

$$\Phi((x_1, \cdots, x_m, y)) = (x_m, y) \quad (x_1, \cdots, x_m, y) \in E^m \times I,$$

and the initial distribution $\tilde{\mu}$ defined by

$$\tilde{\mu} = \mathcal{L}_{P_\mu}(X_1, \cdots, X_{m-1}) \otimes \mu^*.$$

Under the law $\tilde{P}_{\tilde{\mu}}$, $\{Z_n\}_{n\geq 0}$ becomes a Markov chain with the state space $(E^m \times I, \mathcal{E}^m \otimes \mathcal{I})$, transition probability $\tilde{R}$ and initial distribution $\tilde{\mu}$.

Let $\tilde{\theta}$: $\prod_{n=0}^{\infty}(E^m \times I) \longrightarrow \prod_{n=0}^{\infty}(E^m \times I)$ be the shift operator w.r.t. $\{Z_n\}_{n\geq 0}$. In particular, for each $j \geq 0$ and $k \geq 1$

$$\tilde{X}_j \circ \tilde{\theta}^k = \tilde{X}_{j+km}.$$

In view of (2.2) and (2.4), by Lemma 3.1-(ii) of de Acosta (1988) we have

$$\mathcal{L}_{\tilde{P}_{\tilde{\mu}}}(\{\tilde{X}_n\}_{n\geq -(m-1)}) = \mathcal{L}_{P'_{\mu'}}(\{X_n\}_{n\geq -(m-1)}).$$

By the definition of $P'_{\mu'}$, in particular we have

(2.6) $$\mathcal{L}_{\tilde{P}_{\tilde{\mu}}}(\{\tilde{X}_n\}_{n\geq 0}) = \mathcal{L}_{P_{\mu}}(\{X_n\}_{n\geq 0}).$$

For any $n \geq 1$ and bounded measurable function h on $(E \times I) \times \prod_{k=1}^{n}(E^m \times I)$,

$$\tilde{E}_{z_0}\big(h(\Phi(Z_0), Z_1, \cdots, Z_n)\big)$$
$$= \int \tilde{R}(z_0, dz_1) \int \cdots \int \tilde{R}(z_{n-1}, dz_n) h(\Phi(z_0), z_1, \cdots, z_n)$$
$$= \int \tilde{P}(\Phi(z_0), dz_1) \int \cdots \int \tilde{P}(\Phi(z_{n-1}), dz_n) h(\Phi(z_0), z_1, \cdots, z_n).$$

Integrating both sides gives

$$\tilde{E}_{\tilde{\mu}}\big(h(\Phi(Z_0), Z_1, \cdots, Z_n)\big)$$
$$= \int \mu^*\big(d(x_0, y_0)\big) \int \tilde{P}((x_0, y_0), dz_1)$$
$$\times \int \cdots \int \tilde{P}(\Phi(z_{n-1}), dz_n) h(\Phi(z_0), z_1, \cdots, z_n).$$

Hence, for any fixed $\Pi \in (\mathcal{E} \otimes \mathcal{I}) \bigotimes_{n=1}^{\infty}(\mathcal{E}^m \otimes \mathcal{I})$, $\tilde{P}_{z_0}(\Pi)$ and $\tilde{P}_{\tilde{\mu}}(\Pi)$ depend only on $\Phi(z_0) = (x_0, y_0)$ and μ^*, respectively. Since the object we are concerned with is the sequence

(2.7) $$\big\{(\tilde{X}_0, Y_0), (\tilde{X}_{(k-1)m+1}, \cdots, \tilde{X}_{km}, Y_k);\ k \geq 1\big\},$$

to emphasize this very useful fact, we use $\tilde{P}_{(x_0,y_0)}$ and $\tilde{P}_{\mu^*}$ to denote the probability distributions of the sequence in (2.7) under the Markovian laws $\tilde{P}_{z_0}$ and $\tilde{P}_{\tilde{\mu}}$, respectively. More precisely, we define

$$\tilde{P}_{(x_0,y_0)} = \tilde{P}_{z_0}\Big|_{(\mathcal{E}\otimes\mathcal{I})\bigotimes_{n=1}^{\infty}(\mathcal{E}^m\otimes\mathcal{I})},$$

$$\tilde{P}_{\mu^*} = \tilde{P}_{\tilde{\mu}}\Big|_{(\mathcal{E}\otimes\mathcal{I})\bigotimes_{n=1}^{\infty}(\mathcal{E}^m\otimes\mathcal{I})}.$$

The notations $\tilde{E}_{\mu^*}$ and $\tilde{E}_{(x_0,y_0)}$ are also used in the obvious way. When $\mu^*((x_0,y_0))$ is fixed, $\tilde{P}_{\mu^*}$ ($\tilde{P}_{(x_0,y_0)}$) is determined by the Markov kernel $\tilde{P}((x,y),\tilde{\Delta})$. Hence we later use the notation $\tilde{P}((x,y),\tilde{\Delta})$ rather than $\tilde{R}$ for the corresponding transition probability. Under the new notations, (2.6) can be rewritten as

$$\mathcal{L}_{\tilde{P}_{\mu^*}}(\{\tilde{X}_n\}_{n\geq 0}) = \mathcal{L}_{P_\mu}(\{X_n\}_{n\geq 0}).$$

Without loss generality, therefore, we can identify $\{\tilde{X}_n\}_{n\geq 0}$ with $\{X_n\}_{n\geq 0}$. Hence we do this for simplicity. We thus have, under such identification, that

(2.8) $$\mathcal{L}_{\tilde{P}_{\mu^*}}(\{X_n\}_{n\geq 0}) = \mathcal{L}_{P_\mu}(\{X_n\}_{n\geq 0}),$$

(2.9) $$X_j \circ \tilde{\theta}^k = X_{j+km} \quad j \geq 0,\ k \geq 1.$$

Under the law $\tilde{P}_{\mu^*}$, the Markov property appears in the following sense: for any bounded and $(\mathcal{E}\otimes\mathcal{I})\bigotimes_{n=1}^{\infty}(\mathcal{E}^m\otimes\mathcal{I})$-measurable function F on $(E\times I)\times\prod_{n=1}^{\infty}(E^m\times I)$ and any $k\geq 1$,

(2.10) $$\tilde{E}\left[F\circ\tilde{\theta}^k\,\Big|\,\tilde{\mathcal{A}}_k\right] = \tilde{E}_{\Phi(Z_k)}[F] \quad \text{a.s. } \tilde{P}_{\mu^*},$$

where

$$
(2.11) \quad \begin{cases} \tilde{\mathcal{A}}_0 = \sigma\{(X_0, Y_0)\}, \\ \tilde{\mathcal{A}}_k = \sigma\{(X_0, Y_0), \cdots, (X_{(k-1)m+1}, \cdots, X_{km}, Y_k)\} \quad (k \geq 1). \end{cases}
$$

More generally, we have the following strong Markov property: for any stopping time τ (w.r.t. $\{\tilde{\mathcal{A}}_k\}_{k \geq 0}$),

$$
(2.12) \quad \tilde{E}\left[F \circ \tilde{\theta}^\tau \middle| \tilde{\mathcal{A}}_\tau\right] = \tilde{E}_{\Phi(Z_\tau)}[F] \quad \text{a.s. } \tilde{P}_{\mu^*}.
$$

In this way, the Markov chain $\{X_n\}_{n \geq 0}$ is embedded into a larger probability space on which the sequence $\{Y_k\}_{k \geq 0}$ of I-valued random variables are also defined. We call the random sequence in (2.7) (or its probability transition $\tilde{P}$) the **split Markov chain** of $\{X_n\}_{n \geq 0}$ (w.r.t. the minorization (1.4)). More precisely, we call $\tilde{P}$ the split Markov chain of $\{X_n\}_{n \geq 0}$ **in the classical sense** if $m = 1$ and the split Markov chain of $\{X_n\}_{n \geq 0}$ **in the general sense** if $m > 1$.

From (2.4) one can verify that

$$
(2.13) \quad \tilde{P}((x,y), \Phi^{-1}(\bar{A})) = \bar{P}((x,y), \bar{A}) \quad (x,y) \in E \times I, \quad \bar{A} \in \mathcal{E} \otimes \mathcal{I},
$$

where $\bar{P}$ is the Markov kernel on $(E \times I, \mathcal{E} \otimes \mathcal{I})$ given by

$$
(2.14) \quad \begin{aligned} &\bar{P}((x,y), \bar{A}) \\ &= I_0(y)\Big\{Q((1-bI_C)I_{(\cdot)})(x) \otimes \delta_0 + Q(bI_C I_{(\cdot)})(x) \otimes \delta_1\Big\}(\bar{A}) \\ &\quad + I_1(y)\Big\{\nu((1-bI_C)I_{(\cdot)}) \otimes \delta_0 + \nu(bI_C I_{(\cdot)}) \otimes \delta_1\Big\}(\bar{A}) \\ &= I_0(y)Q^*(x, \bar{A}) + I_1(y)\nu^*(\bar{A}) \quad (x,y) \in E \times I, \quad \bar{A} \in \mathcal{E} \otimes \mathcal{I}. \end{aligned}
$$

This means that under the law $\tilde{P}_{\mu^*}$, the sequence $\{\Phi_n\}_{n \geq 0}$ defined by

$$
(2.15) \quad \Phi_n = \Phi(Z_n) = (X_{nm}, Y_n) \quad n = 0, 1, \cdots
$$

becomes a Markov chain with the state space $(E \times I, \mathcal{E} \otimes \mathcal{I})$, the transition probability being $\bar{P}((x,y), \bar{A})$ and the initial distribution μ^*. Clearly, $\{\Phi_n\}_{n \geq 0}$ is nothing but the split Markov chain of the chain $\{X_{nm}\}_{n \geq 0}$ in the classical sense and it coincides with the split chain in (2.7) when $m = 1$. From (2.14) one can verify that π^* is the invariant measure (distribution) of the chain $\{\Phi_n\}_{n \geq 0}$ and that the set α^* given by

$$(2.16) \qquad \alpha^* = C \times \{1\}$$

is an atom of the chain $\{\Phi_n\}_{n \geq 0}$ with

$$(2.17) \qquad \bar{P}(\alpha^*, \cdot) = \nu^*(\cdot).$$

One can easily see from (2.14) that $E \times \{1\}$ is also an atom which was used in the previous literature. The reason why we choose $C \times \{1\}$ is that the smaller set is sometimes easier to control than the bigger one (see the proof of Theorem 3.4 below for an example).

From (2.3),

$$(2.18) \qquad \mu^*(\alpha^*) = b\mu(C)$$

for every measure μ on $(E, \mathcal{E})$. According to Theorem 9.1.6 of Meyn-Tweedie (1993), the chain $\{X_{nm}\}_{n \geq 0}$ is Harris recurrent if the chain $\{X_n\}_{n \geq 0}$ is Harris recurrent and aperiodic, in which case one can see from the relation (2.8) that the chain $\{\Phi_n\}_{n \geq 0}$ is Harris recurrent. We summarize our observations into the following proposition.

Proposition 2.1. *Under the law $\tilde{P}_{\mu^*}$, the random sequence $\{\Phi_n\}_{n \geq 0}$ given in (2.15) becomes a Markov chain with the state space $(E \times I, \mathcal{E} \otimes \mathcal{I})$, the transition probability $\bar{P}((x,y), \bar{A})$ and the initial distribution μ^*. The set*

α^* given by (2.16) is an atom of the chain $\{\Phi_n\}_{n\geq 0}$. The chain $\{\Phi_n\}_{n\geq 0}$ is positive (with the invariant distribution π^*) whenever $\{X_n\}_{n\geq 0}$ has the properties; $\{\Phi_n\}_{n\geq 0}$ is Harris recurrent if $\{X_n\}_{n\geq 0}$ is Harris recurrent and aperiodic.

We call the Markov chain $\{\Phi_n\}_{n\geq 0}$ (or its transition $\bar{P}$) the **marginal** of the split chain given in (2.7). We have thus constructed two Markov chains on the same probability space by the splitting procedure: One is the split chain $\tilde{P}$ with the state space $E^m \times I$ and another is its marginal $\bar{P}$ with the state space $E \times I$. Both of them are so important in this study that all notations introduced for these two chains will be exclusively used without further comment.

We have also created an atom α^* for the marginal chain $\bar{P}$ and, for the split chain $\tilde{P}$ in the case $m = 1$. Although it may not be an atom of $\tilde{P}$ in the general case, α^* still possesses some properties similar to an atom. We now discuss these properties. To make our discussion meaningful, we assume that $\{X_n\}_{n\geq 0}$ is Harris recurrent (and aperiodic if $m > 1$). We define the regeneration times $\{\tau(k)\}_{k\geq 0}$ as follows:

$$(2.19) \quad \begin{cases} \tau(0) = \inf\{n \geq 0;\ \Phi_n \in \alpha^*\} \\ \tau(k+1) = \inf\{n > \tau(k);\ \Phi_n \in \alpha^*\}, \quad (k \geq 0) \end{cases}$$

Clearly, $\{\tau(k)\}_{k\geq 0}$ are stopping times w.r.t. both $\{\Phi_n\}_{n\geq 0}$ and the split chain given in (2.7). By Proposition 2.1, $\tau(k)$ is almost surely finite for each $k \geq 0$.

From (2.13) and (2.17) we have:

$$(2.20) \quad \tilde{P}(\alpha^*, \Phi^{-1}(\bar{A})) = \nu^*(\bar{A}) \qquad \forall \bar{A} \in \mathcal{E} \otimes \mathcal{I}.$$

From (2.10), therefore, for any $(\mathcal{E} \otimes \mathcal{I}) \bigotimes_{n=1}^{\infty} (\mathcal{E}^m \otimes \mathcal{I})$-measurable and bounded function F on $(E \times I) \times \prod_{n=1}^{\infty} (E^m \times I)$ and $(x_0, y_0) \in \alpha^*$,

$$\tilde{E}_{(x_0,y_0)}[F \circ \tilde{\theta}] = \tilde{E}_{(x_0,y_0)}\{\tilde{E}_{\Phi_1}[F]\}$$
$$= \int \nu^*(d(x_1,y_1))\tilde{E}_{(x_1,y_1)}[F] = \tilde{E}_{\nu^*}[F].$$

This observation is written in the form

(2.21) $$\tilde{E}_{\alpha^*}[F \circ \tilde{\theta}] = \tilde{E}_{\nu^*}[F].$$

By using the strong Markov property, which will be exploited repeatedly in this paper, we obtain the following important result:

Theorem 2.2. *Assume that $\{X_n\}_{n\geq 0}$ is Harris recurrent and aperiodic (if $m > 1$) and let $\{n_k\}_{k\geq 0}$ be a sequence of strictly increasing non-negative integers. Then, under the law $\tilde{P}_{\mu^*}$, the following random blocks:*

$$B_0 = (X_0, X_1, \cdots, X_{m\tau(n_0)}),$$
$$B_1 = (X_{m(\tau(n_0)+1)}, X_{m(\tau(n_0)+1)+1}, \cdots, X_{m\tau(n_1)}),$$
$$\cdots\cdots\cdots\cdots\cdots\cdots\cdots\cdots\cdots\cdots\cdots\cdots\cdots$$
$$B_k = (X_{m(\tau(n_{k-1})+1)}, X_{m(\tau(n_{k-1})+1)+1}, \cdots, X_{m\tau(n_k)}),$$
$$\cdots\cdots\cdots\cdots\cdots\cdots\cdots\cdots\cdots\cdots\cdots\cdots\cdots$$

are independent and

$$\mathcal{L}_{\tilde{P}_{\mu^*}}(\{X_{m(\tau(n_{k-1})+1)}, \cdots, X_{m\tau(n_k)}\})$$
$$= \mathcal{L}_{\tilde{P}_{\nu^*}}(\{X_0, \cdots, X_{m\tau(n_k - n_{k-1} - 1)}\}) \quad k = 1, 2, \cdots.$$

The proof of Theorem 2.2 is given in Section 1 of Appendix.

By taking $n_k = k$ in Theorem 2.2 in the case $m = 1$, we have the following corollary:

Corollary 2.3. *Assume that $\{X_n\}_{n\geq 0}$ is Harris recurrent and that $m=1$. Then, under the law $\tilde{P}_{\mu^*}$, the following random blocks:*

$$B_0 = (X_0, \cdots, X_{\tau(0)}),$$
$$B_1 = (X_{\tau(0)+1}, \cdots, X_{\tau(1)}),$$
$$\cdots\cdots\cdots\cdots\cdots\cdots\cdots\cdots\cdots\cdots\cdots\cdots\cdots\cdots\cdots\cdots\cdots\cdots$$
$$B_k = (X_{\tau(k-1)+1}, \cdots, X_{\tau(k)}),$$
$$\cdots\cdots\cdots\cdots\cdots\cdots\cdots\cdots\cdots\cdots\cdots\cdots\cdots\cdots\cdots\cdots\cdots\cdots$$

are independent and among them, $\{B_k\}_{k\geq 1}$ are i.i.d. with the common distribution:

$$\mathcal{L}_{\tilde{P}_{\nu^*}}(\{X_0, \cdots, X_{\tau(0)}\}).$$

Recall that a sequence $\{B_k\}_{k\geq 1}$ of strictly stationary random elememts is called **1-dependent** if for every $k > 1$, the following two collections

$$\{B_1, \cdots, B_{k-1}\} \quad \text{and} \quad \{B_{k+1}, B_{k+1}, \cdots\}$$

are independent. Note that, by definition, $m\tau(k) + m - 1 < m\tau(k+1)$ for each $k \geq 0$. By Theorem 2.2 we have the following:

Corollary 2.4. *Assume that $\{X_n\}_{n\geq 0}$ is Harris recurrent and aperiodic. Then, under the law $\tilde{P}_{\mu^*}$, the random blocks $\{B_k\}_{k\geq 1}$ given by*

$$B_k = (X_{m(\tau(k-1)+1)}, X_{m(\tau(k-1)+1)+1}, \cdots, X_{m\tau(k)+m-1}) \quad k=1,2,\cdots$$

are 1-dependent with the common distribution:

$$\mathcal{L}_{\tilde{P}_{\nu^*}}(\{X_0, X_1, \cdots, X_{m\tau(0)+m-1}\}).$$

The following theorem shows that the ergodicity is preserved through the splitting procedure and furthermore, $\{\Phi_n\}_{n\geq 0}$ and $\{X_n\}_{n\geq 0}$ have equivalent distances between their n-step transition probabilities and their invariant distributions.

Theorem 2.5. *Assume that $\{X_n\}_{n\geq 0}$ is a positive chain. Then,*

(i) If $\{X_n\}_{n\geq 0}$ is ergodic, then so is $\{\Phi_n\}_{n\geq 0}$;

(ii) For every probability measure μ on $(E, \mathcal{E})$ and for every $n \geq 1$

$$(2.22) \qquad \|\mu P^{mn} - \pi\|_V = \|\mu^* \bar{P}^n - \pi^*\|_V;$$

(iii) For every $n \geq 2$,

$$(2.23) \qquad \int \|P^{mn}(x, \cdot) - \pi\|_V \pi(dx) \leq \int \|\bar{P}^n((x,y), \cdot) - \pi^*\|_V \pi^*(dx)$$
$$\leq \int \|P^{m(n-1)}(x, \cdot) - \pi\|_V \pi(dx).$$

The proof of Theorem 2.5 is given in Section 2 of Appendix.

I-3. Maximal and minimal inequalities

Let B be a separable Banach space and let $\xi \colon E \longrightarrow B$ be a measurable map and set

$$S_n = \sum_{k=0}^{n-1} \xi(X_k), \quad n = 1, 2, \cdots.$$

Let $\mathcal{P}(B)$ be the metric space of all probability measures on B endowed with the Prohorov metric ρ, whose metric topology is, as we know (Theorem 5, p.238 of Billingsley (1968)), equivalent to the weak convergence topology on $\mathcal{P}(B)$.

Before the main results of this section, we prove the following:

Proposition 3.1. *Let $\{X_n\}_{n\geq 0}$ be an aperiodic and Harris recurrent Markov chain and let $\{a_n\}$ be a sequence of positive numbers satisfying $a_n \to +\infty$ $(n \to \infty)$. Then for any probability measures μ and λ on $(E, \mathcal{E})$,*

$$\lim_{n\to\infty} \rho\left(\mathcal{L}_{P_\mu}\left(\frac{S_n}{a_n}\right), \mathcal{L}_{P_\lambda}\left(\frac{S_n}{a_n}\right)\right) = 0. \tag{3.1}$$

Proof. By Theorem 18.1.2 of Meyn-Tweedie (1993),

$$\lim_{n\to\infty} \int \|P^n(x,\cdot) - P^n(y,\cdot)\|_V \mu(dx)\lambda(dy) = 0.$$

Given $\epsilon > 0$, therefore, we can choose $N_1 \geq 1$ such that

$$\int \|P^{N_1}(x,\cdot) - P^{N_1}(y,\cdot)\|_V \mu(dx)\lambda(dy) < \frac{\epsilon}{3}.$$

Choose $N_2 \geq 1$ such that for all $n \geq N_2$

$$\max\left\{P_\mu\{\|S_{N_1}\| > \frac{\epsilon a_n}{2}\}, P_\lambda\{\|S_{N_1}\| > \frac{\epsilon a_n}{2}\}\right\} < \frac{\epsilon}{3}.$$

Hence, for any Borel subset A of B and for any $n \geq max\{N_1, N_2\}$

$$P_\mu\left\{\frac{S_n}{a_n} \in A\right\}$$
$$\leq P_\mu\left\{\|S_{N_1}\| > \frac{\epsilon a_n}{2}\right\} + P_\mu\left\{\frac{S_n - S_{N_1}}{a_n} \in A^{\epsilon/2}\right\}$$
$$\leq \frac{\epsilon}{3} + E_\mu P_{X_{N_1}}\left\{\frac{S_{n-N_1}}{a_n} \in A^{\epsilon/2}\right\}$$
$$\leq \frac{\epsilon}{3} + \int \|P^{N_1}(x,\cdot) - P^{N_1}(y,\cdot)\|_V \mu(dx)\lambda(dy) + P_\lambda\left\{\frac{S_n - S_{N_1}}{a_n} \in A^{\epsilon/2}\right\}$$
$$\leq \frac{2\epsilon}{3} + P_\lambda\left\{\|S_{N_1}\| > \frac{\epsilon a_n}{2}\right\} + P_\lambda\left\{\frac{S_n}{a_n} \in A^\epsilon\right\}$$
$$\leq \epsilon + P_\lambda\left\{\frac{S_n}{a_n} \in A^\epsilon\right\},$$

where $A^\delta \equiv \{x \in B;\ d(x, A) \leq \delta\}$ $(\delta > 0)$. Hence, by the definition of the Prohorov metric

$$\rho\left(\mathcal{L}_{P_\mu}\left(\frac{S_n}{a_n}\right), \mathcal{L}_{P_\lambda}\left(\frac{S_n}{a_n}\right)\right) \leq \epsilon \quad n \geq max\{N_1, N_2\}.$$

We have proved (3.1). ∎

Proposition 3.1 tells us that the distributional limiting behavior of the normalized sequence $\{S_n/a_n\}_{n\geq 1}$ is independent of the initial distribution when $\{X_n\}_{n\geq 0}$ is aperiodic and Harris recurrent. In this case we don't need to specify the initial distribution in the statements like

"$\{S_n/a_n\}_{n\geq 1}$ converges in distribution",

"$\{S_n/a_n\}_{n\geq 1}$ is bounded in probability",

or

"$S_n/a_n \longrightarrow 0$ in probability".

In the rest of this section, we define summation, maximum and the minimum over the empty set as follows

$$\sum_{k\in\phi} a_k = 0, \quad \max_{k\in\phi} a_k = 0 \quad \text{and} \quad \min_{k\in\phi} a_k = +\infty,$$

where $\{a_k\}$ is a sequence of non-negative numbers. We let the minorization (1.4) be fixed and we now state our first result on the maximal and minimal inequalities.

Theorem 3.2. *Assume that $\{X_n\}_{n\geq 0}$ is Harris recurrent (and aperiodic if $m > 1$) and define $\{i_C(n)\}_{n\geq 1}$ and $\{\tau_C^m(k)\}_{k\geq 0}$ as follows:*

$$(3.2) \qquad i_C(n) = \sum_{k=1}^{[n/m]-1} I_C(X_{km}) \qquad n = 1, 2, \cdots,$$

$$(3.3) \quad \begin{cases} \tau_C^m(1) = \inf\{n \geq 1; \ X_{mn} \in C\} \\ \tau_C^m(k+1) = \inf\{n > \tau_C^m(k); \ X_{mn} \in C\} \quad (k \geq 1). \end{cases}$$

Then, for every $s, t > 0$, every $x_o \in B$ and every $n \geq 1$,

$$(3.4) \quad P_\mu\{\|S_n - x_o\| < s + t\} \geq br P_\mu\{\min_{k \leq i_C(n)} \|S_{m\tau_C^m(k)+1} - x_o\| < t\},$$

where $r = \min_{1 \leq k \leq n} P_\nu\{\|S_k\| < \frac{s}{2}\} - \frac{1}{b}\sup_{x \in C} P_x\{\sum_{j=1}^{m-1} \|\xi(X_j)\| > \frac{s}{2}\}$ when $m > 1$, and $\min_{1 \leq k \leq n} P_\nu\{\|S_k\| < s\}$ when $m = 1$. In addition,

$$(3.5) \quad P_\mu\{\max_{k \leq i_C(n)} \|S_{m\tau_C^m(k)+1}\| > s + t\} \leq b^{-1}(1-c)^{-1} P_\mu\{\|S_n\| > t\},$$

where $c = \max_{1 \leq k \leq n} P_\nu\{\|S_k\| > \frac{s}{2}\} + \frac{1}{b}\sup_{x \in C} P_x\{\sum_{j=1}^{m-1} \|\xi(X_j)\| > \frac{s}{2}\}$ when $m > 1$, and $\max_{1 \leq k \leq n} P_\nu\{\|S_k\| > s\}$ when $m = 1$, $b > 0$ is as in (1.4), and we assume that $c < 1$.

We now assume that $\{X_n\}_{n \geq 0}$ has an atom α and we specify (1.4) by taking $C = \alpha$, $b = 1$, $m = 1$ and $\nu = P(\alpha, \cdot)$. Immediately, we have the following:

Corollary 3.3. *Assume that $\{X_n\}_{n \geq 0}$ is Harris recurrent and that $\{X_n\}_{n \geq 0}$ has an atom α. Define*

$$(3.6) \quad i_\alpha(n) = \sum_{k=1}^{n-1} I_\alpha(X_k) \quad n = 1, 2, \cdots,$$

$$(3.7) \quad \begin{cases} \tau_\alpha(1) = \tau_\alpha \\ \tau_\alpha(k+1) = \inf\{n > \tau_\alpha(k); \; X_n \in \alpha\} \quad (k \geq 1). \end{cases}$$

Then, for every $s, t > 0$, every $x_o \in B$ and every $n \geq 1$,

$$(3.8) \quad P_\mu\{\|S_n - x_o\| < s + t\} \geq r P_\mu\{\min_{k \leq i_\alpha(n)} \|S_{\tau_\alpha(k)+1} - x_o\| < t\},$$

where

$$r = \min_{1 \leq k \leq n} P_\alpha\{\|\sum_{j=1}^{k} \xi(X_j)\| < s\};$$

and,

(3.9) $\quad P_\mu\{\max_{k \leq i_\alpha(n)} \|S_{\tau_\alpha(k)+1}\| > s+t\} \leq (1-c)^{-1} P_\mu\{\|S_n\| > t\},$

where

$$c = \max_{1 \leq k \leq n} P_\alpha\{\|\sum_{j=1}^{k} \xi(X_j)\| > s\}$$

and we assume that $c < 1$.

We now consider the split chain $\tilde{P}$ generated by the minorization (1.4). Similarly, we have the following:

Theorem 3.4. *Assume that $\{X_n\}_{n \geq 0}$ is Harris recurrent (and aperiodic if $m > 1$) and let $\{\tau(k)\}_{k \geq 0}$ be defined as in (2.19). Define*

(3.10) $\quad i(n) = \sum_{k=0}^{[n/m]-1} I_{\alpha^*}(\Phi_k) \quad n = 1, 2, \cdots,$

Then, for every $s, t > 0$, every $x_o \in B$ and every $n \geq 1$,

(3.11) $\quad P_\mu\{\|S_n - x_o\| < s+t\} \geq r\tilde{P}_{\mu^*}\{\min_{0 \leq k < i(n)} \|S_{m\tau(k)+1} - x_o\| < t\},$

where $r = \min_{1 \leq k \leq n} P_\nu\{\|S_k\| < \frac{s}{2}\} - \frac{1}{b}\sup_{x \in C} P_x\{\sum_{j=1}^{m-1} \|\xi(X_j)\| > \frac{s}{2}\}$ when $m > 1$, and $\min_{1 \leq k \leq n} P_\nu\{\|S_k\| < s\}$ when $m = 1$. Furthermore,

(3.12) $\quad \tilde{P}_{\mu^*}\{\max_{0 \leq k < i(n)} \|S_{m\tau(k)+1}\| > s+t\} \leq (1-c)^{-1} P_\mu\{\|S_n\| > t\},$

where $c = \max_{1 \leq k \leq n} P_\nu\{\|S_k\| > \frac{s}{2}\} + \frac{1}{b}\sup_{x \in C} P_x\{\sum_{j=1}^{m-1} \|\xi(X_j)\| > \frac{s}{2}\}$ when $m > 1$, and $\max_{1 \leq k \leq n} P_\nu\{\|S_k\| > s\}$ when $m = 1$ and, we assume that $c < 1$.

Because of the similarity in proof, only the proof of Theorem 3.4 is given in Section 3 of Appendix. We leave Theorem 3.2 to the reader.

I-4. Maximal integrability on small sets

In the study of strong limit theorems, usually one needs to control the tail probability of $\max_{k\leq n} ||S_k||$. In the independent context, this is carried out by applying a maximal inequality of Lévy-Ottaviani's type. The maximal inequality we establish is the best thing we can do in the Markovian context and unfortunately, it is not strong enough to control the whole trajectory of $\{S_n\}_{n\geq 1}$ (see Example 2.1 in Chapter III for one of the resulting consequences). Based on this consideration, we look at the following type of integrability:

$$(4.1) \qquad \int_C \pi(dx) E_x \max_{n\leq \tau_C} \varphi(||S_n||) < +\infty,$$

where C is a small set and $\varphi: [0,+\infty) \longrightarrow [0,+\infty)$ is a non-decreasing function such that for some fixed $\lambda > 0$

$$(4.2) \qquad \varphi(cs) \leq c^\lambda \varphi(s) \quad \text{for all } s \geq 0 \text{ and } c \geq 2.$$

In the following discussion, let such a function φ and the measurable map $\xi: E \longrightarrow B$ be fixed. To make our discussion meaningful, we assume that $\{X_n\}_{n\geq 0}$ is Harris recurrent. Define

$$\mathcal{S} = \{\text{all small sets}\},$$

$$\mathcal{S}_\varphi(\xi) = \Big\{C \in \mathcal{S}; \ \int_C \pi(dx) E_x \max_{n\leq \tau_C} \varphi(||S_n||) < +\infty\Big\}.$$

Our first goal in this section is to prove the following:

Theorem 4.1. *If the Markov chain $\{X_n\}_{n\geq 0}$ is aperiodic and Harris recurrent then,*

$$(4.3) \qquad \mathcal{S}_\varphi(\xi) = \phi \quad \text{or} \quad \mathcal{S}.$$

As far as we know, a property like (4.3) for Markov chains was never mentioned before in the literature. In chapter III, we shall use it to describe some strong limit theorems for ergodic Markov chains. We can also find some interesting applications to classical Markov chain theory by specifying B, ξ and φ. Taking, for example, $B = \mathbf{R}$ and $\xi \equiv 1$ in Theorem 4.1, immediately we conclude that the recurrence condition defined by

$$\int_C \pi(dx) E_x \varphi(\tau_C) < +\infty$$

either holds for all $C \in \mathcal{S}$ or for none of them. More generally, if we take $B = \mathbf{R}$ and use $|\xi|$ instead of ξ, then we conclude that the regularity condition defined by

$$\int_C \pi(dx) E_x \varphi\Big(\sum_{k=0}^{\tau_C - 1} |\xi(X_k)| \Big) < +\infty$$

either holds for all $C \in \mathcal{S}$ or for none of them.

The proof of Theorem 4.1 is given in Section 4 of Appendix. Although the topics are completely different, some ideas we use in our argument come from Klass ([28], [29]).

We now assume that $\{X_n\}_{n \geq 0}$ is aperiodic and Harris recurrent and,

(4.4) $$\mathcal{S}_\varphi(\xi) = \mathcal{S}.$$

For the purpose of applications, we need to build up a split chain to which the integrability we discuss is inheritable. Let $C \in \mathcal{S}$ be fixed. By definition we can find $0 < b < 1/2$, $m \geq 1$ and a probability measure ν on $(E, \mathcal{E})$ such that

(4.5) $$P^m \geq 2b I_C \otimes \nu.$$

In particular,

(4.6) $$P^m \geq b I_C \otimes \nu.$$

Theorem 4.2. *Let $\{X_n\}_{n\geq 0}$ be an aperiodic Harris recurrent Markov chain and assume (4.4) holds. Given $C \in \mathcal{S}$ with (4.5), let $\tilde{P}$ be the split chain generated by the minorization (4.6). Then,*

$$(4.7) \qquad \tilde{E}_{\nu^*} \max_{n \leq m(\tau(0)+1)} \varphi(\|S_n\|) < +\infty,$$

where $\tau(0)$ is given in (2.19).

Chapter II. The Central Limit Theorem

II-1. Introduction

In this chapter, $\{X_n\}_{n\geq 0}$ is an ergodic Markov chain with invariant distribution π and $\xi: E \longrightarrow \mathbf{R}$ is a measurable function. Write

$$S_n = \sum_{k=0}^{n-1} \xi(X_k), \quad n = 1, 2, \cdots.$$

The central limit theorem (CLT) states that

$$S_n/\sqrt{n} \longrightarrow N(0, \sigma^2) \text{ in distribution,}$$

where $\sigma^2 \geq 0$ and we define $N(0, 0^2) = \delta_0$. According to Proposition I-3.1, this kind of limit behavior does not depend on the initial distribution of the Markov chain. In addition to its importance in probability theory, the CLT in Markovian context plays a crucial role in describing and identifying some other limit theorems for Markov chains such as the law of the iterated logarithm (Chapter III) and the moderate deviation principle (Chapter IV).

The CLT for ergodic Markov chains has been under study for years and an extensive literature exists (e.g. see [4], [17], [18], [32], [33], [37], [39]). It is apparent now that the conditions

$$(1.1) \quad \int \xi(x)\pi(dx) = 0 \text{ and } \int \xi^2(x)\pi(dx) < +\infty$$

are no longer sufficient for the CLT in the Markovian context. The following simple example shows that they aren't necessary either:

Example 1.1. Let $\{Z_n\}_{n\geq 0}$ be a sequence of i.i.d. real random variables (note that no integrability condition is assumed). Then, $\{(Z_n, Z_{n+1})\}_{n\geq 0}$

is an ergodic Markov chain with the state space $\mathbf{R}^2$ (In fact, we can see that $\{(Z_n, Z_{n+1})\}_{n\geq 0}$ has the same type of transition as the one defined in Example I-1.1). Define $\xi: \mathbf{R}^2 \longrightarrow \mathbf{R}$ by $\xi(x,y) = x - y$, $x,y \in \mathbf{R}$. One can see that

$$S_n/\sqrt{n} \longrightarrow 0 \text{ in probability.}$$

Hence the (degenerate) CLT holds.

This example also suggests that it seems impossible to identify the limit variance σ^2 with a single formula under conditions which are merely enough for the validitying of the CLT. In other words, the validity of the CLT and the identification of the limiting variance are two different problems in the Markovian context.

As far as we know, all the results concerning the first problem are basically of the following two types: Assume that (Niemi-Nummelin (1982) and Nummelin (1984))

$$(1.2) \quad \begin{cases} \int \xi(x)\pi(dx) = 0 \text{ and } \int \xi^2(x)\pi(dx) < +\infty, \\ \int_A \pi(dx) E_x\left(\sum_{k=1}^{\tau_A} |\xi(X_k)|\right)^2 < +\infty \quad \forall A \in \mathcal{E}^+ \end{cases}$$

or that (Maigret (1978))

$$(1.3) \quad \begin{cases} \int \xi(x)\pi(dx) = 0 \text{ and } \int \xi^2(x)\pi(dx) < +\infty, \\ \text{There exists } l \in L^2(\pi) \text{ such that } l - Pl = \xi. \end{cases}$$

Then, the CLT holds. Two examples given in Niemi-Nummelin (1982) show that the conditions (1.2) and (1.3) are incomparable.

Concerning the second problem, de Acosta ([4]) recently proved that

$$(1.4) \quad \sigma^2 = \int \xi^2(x)\pi(dx) + 2\int \sum_{n=1}^{\infty} \xi(x) P^n \xi(x) \pi(dx)$$

if $\{X_n\}_{n\geq 0}$ is ergodic of degree 2 and ξ is bounded with $\int \xi(x)\pi(dx) = 0$, in which case the series $\sum_{n=1}^{\infty} \xi(\cdot)P^n\xi(\cdot)$ converges in $L^1(\pi)$. The representation (1.4) appears to be a natural and reasonable solution to the second problem.

What we intend to do in this chapter is the following: In Section II-2 we prove that the CLT is essentially implied by the stochastic boundedness of $S_n/\sqrt{n}$, which is trivially necessary for the CLT. This also means that the normal distribution is the only possible limit distribution for the normalized sum $S_n/\sqrt{n}$. In Section II-3, we obtain (1.4) in a more general setting. Roughly speaking, we prove that (1.4) holds as soon as the right-hand side of (1.4) makes sense. In Section II-4 we prove that ergodicity of degree 2 is the weakest ergodicity condition which allows the CLT to hold for all bounded and mean zero function ξ. This result gives an interpretation of ergodicity of degree 2 which is in a sense more natural than the standard definition used in Nummelin (1984). As an application of our main theorem, we prove that uniform ergodicity is strong enough to ensure that the CLT holds for all ξ satisfying (1.1).

The split and regeneration methods established in Chapter I are fundamental to our proof. The argument we follow also relies on the Wald's equations and the resolvent approximation of Markov transitions, the method previously used in de Acosta (1988) and further developed in this chapter.

II-2. Validity of the CLT

We begin with a simpler situation: chains with an atom.

Theorem 2.1. *Let $\{X_n\}_{n\geq 0}$ be an ergodic Markov chain with an atom α. Then, the following three statements are equivalent:*

(i) $S_n/\sqrt{n} \longrightarrow N(0,\sigma^2)$ in distribution for some $\sigma^2 \geq 0$;

(ii) $\{S_n/\sqrt{n}\}_{n\geq 1}$ is bounded in probability;

(iii) $E_\alpha\left(\sum_{k=1}^{\tau_\alpha}\xi(X_k)\right) = 0$ and $E_\alpha\left(\sum_{k=1}^{\tau_\alpha}\xi(X_k)\right)^2 < +\infty.$

Furthermore,

$$(2.1) \qquad \sigma^2 = \pi(\alpha) \cdot E_\alpha\left(\sum_{k=1}^{\tau_\alpha}\xi(X_k)\right)^2$$

whenever these equivalent statements hold.

Proof. The direction "(i)$\Longrightarrow$(ii)" is trivial. We need only to prove "(ii)$\Longrightarrow$(iii)" and "(iii)$\Longrightarrow$(i)".

Let $\{\tau_\alpha(k)\}_{k\geq 1}$ and $i_\alpha(n)$ be defined as in Corollary I-3.3. and define

$$(2.2) \qquad \xi_k = \sum_{j=\tau_\alpha(k)+1}^{\tau_\alpha(k+1)} \xi(X_j) \qquad (k \geq 1).$$

Then $\{\xi_k\}_{k\geq 1}$ are i.i.d. random variables with the common distribution:

$$\mathcal{L}_{P_\alpha}\left(\sum_{k=1}^{\tau_\alpha}\xi(X_k)\right).$$

From (ii), for sufficiently large $M > 0$ we have

$$\sup_n P_\alpha\{|\sum_{k=1}^n \xi(X_k)| > M\sqrt{n}\} \leq \frac{1}{2}.$$

Let $0 < \lambda < \pi(\alpha)$ be fixed and take $B = \mathbf{R}$ in (I-3.9) of Corollary I-3.3. Then,

$$P_\mu\{|S_{\tau_\alpha([n\lambda])+1}| > 2M\sqrt{n}, \ i_\alpha(n) \geq [n\lambda]\}$$
$$\leq P_\mu\{\max_{k\leq i_\alpha(n)} |S_{\tau_\alpha(k)+1}| > 2M\sqrt{n}\}$$
$$\leq 2P_\mu\{|S_n| > M\sqrt{n}\}.$$

By the law of the large numbers for Markov chains (Theorem 17.1.7 of Meyn-Tweedie (1993)),

$$(2.3) \qquad \lim_{n\to\infty} \frac{i_\alpha(n)}{n} = \pi(\alpha) \qquad \text{a.s.}$$

In particular we have

$$P_\mu\{i_\alpha(n) \geq [n\lambda]\} \longrightarrow 1 \quad (n \to \infty).$$

Therefore the statement (ii) implies that $\{S_{\tau_\alpha([n\lambda])+1}/\sqrt{n}\}_{n\geq 1}$ is bounded in probability. Note that

$$S_{\tau_\alpha([n\lambda])+1} = S_{\tau_\alpha(1)+1} + \sum_{k=1}^{[n\lambda]-1} \xi_k \quad (n \geq 1).$$

Therefore, one can easily see that $\{\sum_{k=1}^n \xi_k/\sqrt{n}\}_{n\geq 1}$ is bounded in probability. By the converse of the central limit theorem for real i.i.d. sequences (we refer to p.273 of Ledoux-Talagrand (1991) for a simple proof), we must have (iii).

To prove "(iii)$\Longrightarrow$(i)" we follow Nummelin (1984), where it is shown how to approximate S_n by the sequence $\{\xi_k\}_{k\geq 1}$. Since

$$(2.4) \quad E\xi_1 = E_\alpha\Big(\sum_{k=1}^{\tau_\alpha} \xi(X_k)\Big) = 0 \quad \text{and} \quad E\xi_1^2 = E_\alpha\Big(\sum_{k=1}^{\tau_\alpha} \xi(X_k)\Big)^2 < +\infty,$$

applying the central limit theorem to the i.i.d. sequence $\{\xi_k\}_{k\geq 1}$ yields

$$(2.5) \quad \frac{1}{\sqrt{n}}\sum_{k=1}^n \xi_k \longrightarrow N(0, E\xi_1^2) \quad \text{in distribution.}$$

Define

$$l_\alpha(n) = \tau_\alpha(i_\alpha(n) \vee 1)$$
$$= \begin{cases} \max\{k \geq 1;\ k \leq n-1 \text{ and } X_n \in \alpha\} & \text{if } \tau_\alpha \leq n-1 \\ \tau_\alpha & \text{if } \tau_\alpha > n-1. \end{cases}$$

Then,

$$(2.6) \quad S_n = S_{n\wedge(\tau_\alpha+1)} + \sum_{k=1}^{i_\alpha(n)-1} \xi_k + \sum_{j=l_\alpha(n)+1}^{n-1} \xi(X_j) \quad (n \geq 1),$$

where $\sum_{k=1}^{i_\alpha(n)-1} \xi_k$ is interpreted to be 0 if $i_\alpha(n) = 0$ or 1 and similarly, $\sum_{j=l_\alpha(n)+1}^{n-1} \xi(X_j) \equiv 0$ if $l_\alpha(n) + 1 > n - 1$. For the first term we have

$$(2.7) \quad P_\mu\{|S_{n\wedge(\tau_\alpha+1)}| > t\} \leq P_\mu\{\sum_{j=1}^{\tau_\alpha} |\xi(X_j)| > t\} \longrightarrow 0 \quad (t \to \infty).$$

For the third term we have

$$P_\mu\{|\sum_{j=l_\alpha(n)+1}^{n-1} \xi(X_j)| > t\}$$

$$\leq \sum_{k=1}^{n-1} P_\mu\{\sum_{j=l_\alpha(n)+1}^{n-1} |\xi(X_j)| > t, \ l_\alpha(n) = n-k\}$$

$$= \sum_{k=1}^{n-1} P_\mu\{\sum_{j=n-k+1}^{n-1} |\xi(X_j)| > t,$$

$$X_{n-k} \in \alpha, \ X_{n-k+1} \notin \alpha, \cdots, X_{n-1} \notin \alpha\}$$

$$(2.8) \quad = \sum_{k=1}^{n-1} P_\mu\{X_{n-k} \in \alpha\}$$

$$\times P_\alpha\{\sum_{j=1}^{k-1} |\xi(X_j)| > t, \ X_1 \notin \alpha, \cdots, X_{k-1} \notin \alpha\}$$

$$= \sum_{k=1}^{n-1} P_\mu\{X_{n-k} \in \alpha\} \cdot P_\alpha\{\sum_{j=1}^{k-1} |\xi(X_j)| > t, \ \tau_\alpha \geq k\}$$

$$\leq \sum_{k=1}^{n-1} P_\alpha\{\sum_{j=1}^{\tau_\alpha} |\xi(X_j)| > t, \ \tau_\alpha \geq k\}$$

$$\leq \sum_{k=1}^{\infty} P_\alpha\{\sum_{j=1}^{\tau_\alpha} |\xi(X_j)| > t, \ \tau_\alpha \geq k\} \longrightarrow 0 \quad (t \to \infty),$$

where the second and the third equalities follow from the Markov property and, the last step follows from the dominated covergence theorem and the fact (Corollary 5.3 of Nummelin(1984)):

$$\sum_{k=1}^{\infty} P_\alpha\{\tau_\alpha \geq k\} = E_\alpha \tau_\alpha < +\infty.$$

Let $e(n) = [n\pi(\alpha)]$ $n = 1, 2, \cdots$. For given $\epsilon > 0$, take

$$\beta = \frac{\epsilon^3}{4(1 + E\xi_1^2)}.$$

Then

$$P_\mu\{|\sum_{k=1}^{i_\alpha(n)-1} \xi_k - \sum_{k=1}^{e(n)} \xi_k| \geq \sqrt{n}\epsilon\}$$

$$\leq 2P\{\max_{k \leq 2n\beta} |\sum_{j=1}^{k} \xi_j| \geq \sqrt{n}\epsilon\} + P_\mu\{|i_\alpha(n) - 1 - e(n)| \geq n\beta\}.$$

By Kolmogorov's inequality

$$P\{\max_{k \leq 2n\beta} |\sum_{j=1}^{k} \xi_j| \geq \sqrt{n}\epsilon\} \leq \frac{1}{\epsilon^2 n} E\left(\sum_{k=1}^{[2n\beta]} \xi_k\right)^2 = \frac{1}{\epsilon^2 n} \cdot [2n\beta]E\xi_1^2 \leq \epsilon.$$

From (2.3)

$$P_\mu\{|i_\alpha(n) - e(n) - 1| \geq n\beta\} \longrightarrow 0 \quad (n \to \infty).$$

Hence, we have proven

(2.9) $$\frac{1}{\sqrt{n}}\left(\sum_{k=1}^{i_\alpha(n)-1} \xi_k - \sum_{k=1}^{e(n)} \xi_k\right) \longrightarrow 0 \quad \text{in probability.}$$

From (2.5) we have

(2.10) $$\frac{1}{\sqrt{n}} \sum_{k=1}^{e(n)} \xi_k \longrightarrow N(0, \sigma^2) \quad \text{in distribution,}$$

where σ^2 is defined by (2.1). In view of (2.9) and (2.10), we get

(2.11) $$\frac{1}{\sqrt{n}} \sum_{k=1}^{i_\alpha(n)-1} \xi_k \longrightarrow N(0, \sigma^2) \quad \text{in distribution.}$$

Therefore, the statement (i) follows from (2.6), (2.7), (2.8) and (2.11). ∎

We now assume $\{X_n\}_{n\geq 0}$ has an order 1 small set, i.e.

(2.12) $$P \geq b I_C \otimes \nu$$

holds for some $0 < b < 1$, some $C \in \mathcal{E}^+$, and some probability measure ν on $(E, \mathcal{E})$. Consider the split chain $\tilde{P}$ generated by the minorization (2.12). Under $\tilde{P}$, $\{X_n\}_{n\geq 0}$ is embedded into a larger probability space together with a sequence $\{Y_n\}_{n\geq 0}$ of $\{0,1\}$ valued random variables, such that $\{(X_n, Y_n)\}_{n\geq 0}$ becomes an ergodic Markov chain (Theorem I-2.5-(i)) with the transition $\tilde{P}$ given in (I-2.4) (with $m = 1$). By (I-2.17), $\tilde{P}(\alpha^*, \cdot) = \nu^*$. Immediately, Theorem 2.1 implies the followng:

Theorem 2.2. *Let $\{X_n\}_{n\geq 0}$ be an ergodic Markov chain satisfying the minorization (2.12). Then, the following three statements are equivalent:*

(i) $S_n/\sqrt{n} \longrightarrow N(0, \sigma^2)$ in distribution for some $\sigma^2 \geq 0$;

(ii) $\{S_n/\sqrt{n}\}_{n\geq 1}$ is bounded in probability;

(iii) $\tilde{E}_{\nu^*}\left(\sum_{k=0}^{\tau(0)} \xi(X_k)\right) = 0$ and $\tilde{E}_{\nu^*}\left(\sum_{k=0}^{\tau(0)} \xi(X_k)\right)^2 < +\infty$,

where $\tau(0)$ is given in (I-2.19).

Further,

(2.13) $$\sigma^2 = \pi^*(\alpha^*) \cdot \tilde{E}_{\nu^*}\left(\sum_{k=0}^{\tau(0)} \xi(X_k)\right)^2$$

whenever these equivalent statements hold.

In the general case, a small set always exists (Theorem 5.2.2 of Meyn-Tweedie (1993)), i.e.

(2.14) $$P^m \geq b I_C \otimes \nu$$

holds for $m \geq 1$, some $0 < b < 1$, some $C \in \mathcal{E}^+$, and some probability measure ν on $(E, \mathcal{E})$. Similarly, we consider the split chain $\tilde{P}$ (possibly in the general sense) generated by the minorization (2.14).

Theorem 2.3. *Let $\{X_n\}_{n \geq 0}$ be an ergodic Markov chain and assume that*

$$\text{(2.15)} \qquad \int \xi^2(x)\pi(dx) < +\infty.$$

Then, the following three statements are equivalent:

(i) $S_n/\sqrt{n} \longrightarrow N(0, \sigma^2)$ in distribution for some $\sigma^2 \geq 0$;

(ii) $\{S_n/\sqrt{n}\}_{n \geq 1}$ is bounded in probability;

(iii) For some (all) minorization (2.14),

$$\tilde{E}_{\nu^*}\left(\sum_{k=0}^{m\tau(0)+m-1} \xi(X_k)\right) = 0 \quad \text{and} \quad \tilde{E}_{\nu^*}\left(\sum_{k=0}^{m\tau(0)+m-1} \xi(X_k)\right)^2 < +\infty.$$

and, when one of these equivalent statements holds,

$$\text{(2.16)} \qquad \begin{aligned} \sigma^2 = & m^{-1}\pi^*(\alpha^*) \cdot \Big\{\tilde{E}_{\nu^*}\Big(\sum_{j=0}^{m\tau(0)+m-1} \xi(X_j)\Big)^2 \\ & + 2\tilde{E}_{\nu^*}\Big(\sum_{j=0}^{m\tau(0)+m-1} \xi(X_j)\Big)\Big(\sum_{k=m(\tau(0)+1)}^{m\tau(1)+m-1} \xi(X_k)\Big)\Big\}, \end{aligned}$$

where $\tau(0)$ and $\tau(1)$ are given in (I-2.19).

Let us make some comments before the proof of Theorem 2.3. By the independence described in Corollary I-2.3, the second term on the right-hand side of (2.16) vanishes when $m = 1$. Therefore (2.16) coincides with (2.13) if $m = 1$. Since they are given in terms of the split chain, (2.13) and (2.16) seem not to be so useful as far as the CLT is concerned. It is technically possible for us to state Theorem 2.2 and 2.3 in such a way that no split chain is involved. However, the representations (2.13) and (2.16) are important in the later development of this work where we use them to connect the CLT with some other limit theorems. Since of course (ii) is necessary for the weak convergence of $\{S_n/\sqrt{n}\}_{n\geq 1}$, these theorems tell us that the normal distribution is the only type of limit distribution for $\{S_n/\sqrt{n}\}_{n\geq 1}$. Besides, statement (ii) is easy to verify sometimes (see the first part of Theorem 3.1 in the next section as an example) so it can be used to establish the CLT in some cases.

Proof of Theorem 2.3. The direction "(i) $\Longrightarrow$ (ii)" is trivial. The proof for "(iii) $\Longrightarrow$ (i)" is almost the same as its counterpart in Theorem 2.1, except that we use the 1-dependent interblocks in Corollary I-2.3 instead of $\{\xi_k\}_{k\geq 1}$ given in (2.2), and the CLT for 1-dependent sequences instead of the CLT for i.i.d. sequences (we refer to the proof of Theorem 7.6 in Nummelin (1984) for the details). We only prove "(ii) $\Longrightarrow$ (iii)". Because of Theorem 2.2, we may assume that $m > 1$ in (2.14). Let the minorization (2.14) be fixed. We use all notations given in Section I-2 without further comment.

Define $\bar{\xi}\colon E \times I \longrightarrow \mathbf{R}$ by $\bar{\xi}(x,y) = \tilde{E}_{(x,y)} \sum_{j=0}^{m-1} \xi(X_j)$, where $(x,y) \in E \times I$. Then by orthogonality we have that

$$\tilde{E}_{\pi^*}\Big(S_{nm} - \sum_{k=0}^{n-1} \bar{\xi}(\Phi_k)\Big)^2$$
$$= \tilde{E}_{\pi^*}\Big(\sum_{k=0}^{n-1}\Big(\sum_{j=0}^{m-1} \xi(X_{km+j}) - \bar{\xi}(\Phi_k)\Big)\Big)^2 = n\tilde{E}_{\pi^*}\Big(\sum_{j=0}^{m-1} \xi(X_j) - \bar{\xi}(\Phi_0)\Big)^2$$
$$\leq n\tilde{E}_{\pi^*}\Big(\sum_{j=0}^{m-1} \xi(X_j)\Big)^2 = nE_\pi\Big(\sum_{j=0}^{m-1} \xi(X_j)\Big)^2,$$

where the last equality follows from (I-2.8). Therefore, (ii) implies that

$$\left\{\sum_{k=0}^{n-1} \bar{\xi}(\Phi_k)/\sqrt{n}\right\}_{n\geq 1} \text{ is bounded in probability.}$$

Note that $\{\Phi_n\}_{n\geq 0}$ is an ergodic Markov chain (Theorem I-2.5) with an atom α^* (Proposition I-2.1) and take (I-2.17) into account. Applying Theorem 2.1 to $\{\Phi_n\}_{n\geq 0}$ gives

(2.17) $$\tilde{E}_{\nu^*}\left(\sum_{k=0}^{\tau(0)} \bar{\xi}(\Phi_k)\right) = 0 \text{ and } \tilde{E}_{\nu^*}\left(\sum_{k=0}^{\tau(0)} \bar{\xi}(\Phi_k)\right)^2 < +\infty.$$

By the Markov property we have

$$\tilde{E}_{\nu^*}\left(\sum_{k=0}^{m\tau(0)+m-1} \xi(X_k)\right) = \tilde{E}_{\nu^*}\left(\sum_{k=0}^{\tau(0)}\sum_{j=0}^{m-1} \xi(X_{km+j})\right)$$

$$= \sum_{k=0}^{\infty} \tilde{E}_{\nu^*}\left(I_{\{\tau(0)\geq k\}} \sum_{j=0}^{m-1} \xi(X_{km+j})\right) = \sum_{k=0}^{\infty} \tilde{E}_{\nu^*}\left(I_{\{\tau(0)\geq k\}} \tilde{E}_{\Phi_k} \sum_{j=0}^{m-1} \xi(X_j)\right)$$

$$= \sum_{k=0}^{\infty} \tilde{E}_{\nu^*}\left(I_{\{\tau(0)\geq k\}} \bar{\xi}(\Phi_k)\right) = \tilde{E}_{\nu^*}\left(\sum_{k=0}^{\tau(0)} \bar{\xi}(\Phi_k)\right) = 0.$$

By orthogonality and the Markov property,

(2.18)
$$\tilde{E}_{\nu^*}\left(\sum_{j=0}^{m\tau(0)+m-1} \xi(X_j) - \sum_{k=0}^{\tau(0)} \bar{\xi}(\Phi_k)\right)^2$$

$$= \tilde{E}_{\nu^*}\left(\sum_{k=0}^{\tau(0)}\left(\sum_{j=0}^{m-1} \xi(X_{km+j}) - \bar{\xi}(\Phi_k)\right)\right)^2$$

$$= \tilde{E}_{\nu^*} \sum_{k=0}^{\tau(0)}\left(\sum_{j=0}^{m-1} \xi(X_{km+j}) - \bar{\xi}(\Phi_k)\right)^2$$

$$= \tilde{E}_{\nu^*} \sum_{k=0}^{\tau(0)} \tilde{E}_{\Phi_k}\left(\sum_{j=0}^{m-1} \xi(X_j) - \bar{\xi}(\Phi_0)\right)^2.$$

Let

(2.19) $$\tau_\alpha^* \equiv \{n \geq 1; \quad \Phi_n \in \alpha^*\}.$$

By the Markov property again,

$$\tilde{E}_{\nu^*}\Big(\sum_{j=0}^{m\tau(0)+m-1} \xi(X_j) - \sum_{k=0}^{\tau(0)} \bar{\xi}(\Phi_k)\Big)^2$$

$$= \tilde{E}_{\alpha^*} \sum_{k=1}^{\tau_\alpha^*} \tilde{E}_{\Phi_k}\Big(\sum_{j=0}^{m-1} \xi(X_j) - \bar{\xi}(\Phi_0)\Big)^2$$

$$= \pi^*(\alpha^*)^{-1}\tilde{E}_{\pi^*}\Big(\sum_{j=0}^{m-1} \xi(X_j) - \bar{\xi}(\Phi_0)\Big)^2$$

$$\leq \pi^*(\alpha^*)^{-1} E_\pi \Big(\sum_{j=0}^{m-1} \xi(X_j)\Big)^2 < +\infty,$$

where the second equality follows from Theorem 10.4.9 of Meyn-Tweedie (1993). By (2.17) we have

$$\tilde{E}_{\nu^*}\Big(\sum_{k=0}^{m\tau(0)+m-1} \xi(X_k)\Big)^2 < +\infty.$$

We have proved (iii). ∎

II-3. Identification of the limiting variance in the CLT

The following is our main result in this section.

Theorem 3.1. *Let $\{X_n\}_{n\geq 0}$ be an ergodic Markov chain and assume that*

(i) $$\int \xi^2(x)\pi(dx) < +\infty;$$

(ii) $$\sum_{n=1}^\infty \int \xi(x) P^n \xi(x) \pi(dx) \text{ converges.}$$

Then,

(3.1) $$S_n/\sqrt{n} \longrightarrow N(0, \sigma^2) \quad \text{in distribution}$$

holds for some $\sigma^2 \geq 0$. Further,

(3.2) $$\sigma^2 = \int \xi^2(x)\pi(dx) + 2\int \sum_{n=1}^{\infty} \xi(x)P^n\xi(x)\pi(dx)$$

if (i) holds and (ii) is strengthened into

(ii') $\qquad\qquad \sum_{n=1}^{\infty} \xi(\cdot)P^n\xi(\cdot)$ converges in $L^1(\pi)$.

To prove this theorem, we need the following observation which links the chain $\{X_n\}_{n \geq 0}$ with its resolvent.

Let $0 < t < 1$ be fixed and consider an i.i.d. Bernoulli sequence $\{\delta_k\}_{k \geq 1}$ with common law given by

(3.3) $$P\{\delta_1 = 0\} = t \quad \text{and} \quad P\{\delta_1 = 1\} = 1 - t.$$

We always assume that $\{\delta_k\}_{k \geq 1}$ is independent of $\{X_n\}_{n \geq 0}$. Define

(3.4) $$\begin{cases} T_0 = 0 \\ T_k = \inf\{n > T_{k-1};\ \delta_n = 1\} \quad (k \geq 1). \end{cases}$$

One can easily verify that $\{T_k - T_{k-1}\}_{k \geq 1}$ is an i.i.d. sequence with the common distribution given by

(3.5) $$P\{T_1 = k\} = (1-t)t^{k-1} \quad k = 1, 2, \cdots.$$

Define the transition P_t on $(E, \mathcal{E})$ as follows:

(3.6) $$P_t(x, A) = (1-t)\sum_{k=1}^{\infty} t^{k-1}P^k(x, A)$$
$$= \sum_{k=1}^{\infty} P\{T_1 = k\}P^k(x, A) \quad x \in E,\ A \in \mathcal{E}.$$

Let $\{X_n^t\}_{n\geq 0}$ be a Markov chain with the transition P_t. We call P_t a **resolvent kernel** of P and call $\{X_n^t\}_{n\geq 0}$ a **resolvent chain** of $\{X_n\}_{n\geq 0}$.

Lemma 3.2. *If $\{X_n\}_{n\geq 0}$ is ergodic then $\{X_n^t\}_{n\geq 0}$ is ergodic with the same invariant distribution π.*

Proof. One can verify directly that π is the invariant distribution of $\{X_n^t\}_{n\geq 0}$. By the generalized Chapman-Kolmogorov equation ((5.46) in Meyn-Tweedie (1993)),

$$(3.7) \qquad P_t^n = \sum_{k=1}^{\infty} P\{T_n = k\} P^k \quad \forall n \geq 1.$$

¿From the fact that

$$(3.8) \qquad P\{T_n = k\} = 0 \quad \forall k < n,$$

one can easily verify by Proposition 6.3 in Nummelin (1984) that

$$\lim_{n\to\infty} \|P_t^n(x,\cdot) - \pi\|_V = 0 \quad \forall x \in E.$$

Hence $\{X_n^t\}_{n\geq 0}$ is ergodic. ∎

Lemma 3.3. *With the same initial distribution for the Markov chains $\{X_n\}_{n\geq 0}$ and $\{X_n^t\}_{n\geq 0}$,*

$$(3.9) \qquad \mathcal{L}\Big(\Big\{\sum_{k=1}^{n} \xi(X_k^t)\Big\}_{n\geq 1}\Big) = \mathcal{L}\Big(\Big\{\sum_{k=1}^{T_n} \delta_k \xi(X_k)\Big\}_{n\geq 1}\Big).$$

Proof. Let μ be the common initial distribution of $\{X_n\}_{n\geq 0}$ and $\{X_n^t\}_{n\geq 0}$. By (5.9) in de Acosta (1988) we have

$$\mathcal{L}(\{X_n^t\}_{n\geq 0}) = \mathcal{L}(\{X_{T_n}\}_{n\geq 0}).$$

Hence, the lemma follows from the following observation:

$$\sum_{k=1}^{n} \xi(X_{T_k}) = \sum_{k=1}^{T_n} \delta_k \xi(X_k).$$

∎

As an application of Lemma 3.3, we have the following:

Lemma 3.4. *Let $\{X_n\}_{n\geq 0}$ be an ergodic Markov chain such that*

(3.10) $$\int \xi^2(x)\pi(dx) < +\infty.$$

Then, the following two statements are equivalent:

(1) $S_n/\sqrt{n} \longrightarrow N(0, \sigma^2)$ in distribution for some $\sigma^2 \geq 0$;

(2) for some (all) $0 < t < 1$, there exists $\sigma_t^2 \geq 0$ such that

$$\sum_{k=0}^{n-1} \xi(X_k^t)/\sqrt{n} \longrightarrow N(0, \sigma_t^2) \quad \text{in distribution.}$$

Further,

(3.11) $$\lim_{t\to 0^+} \sigma_t^2 = \sigma^2$$

whenever these two equivalent statements hold.

Proof. By Proposition I-3.1 we may choose π as the initial distribution for both $\{X_n\}_{n\geq 0}$ and $\{X_n^t\}_{n\geq 0}$. Let

$$e(n) = [(1-t)^{-1}n] \quad n = 1, 2, \cdots.$$

First we claim that under either one of the statements (1) and (2),

(3.12) $$(S_{T_n} - S_{e(n)})/\sqrt{n} \longrightarrow 0 \quad \text{in probability.}$$

Given $\epsilon > 0$, by the CLT for i.i.d. sequences there exists $M > 0$ such that
$$P\{|T_n - e(n)| > M\sqrt{n}\} < \epsilon$$
for all $n \geq 1$. By the independence between $\{T_n\}_{n\geq 0}$ and $\{X_n\}_{n\geq 0}$,
$$P\{|S_{T_n} - S_{e(n)}| > \epsilon\sqrt{n}\}$$
$$\leq \epsilon + \sum_{|k-e(n)|\leq M\sqrt{n}} P\{T_n = k\}P_\pi\{|S_{|k-e(n)|}| > \epsilon\sqrt{n}\}.$$

The law of the large numbers for the ergodic Markov chains (Theorem 17.1.7 in Meyn-Tweedie (1993)) states that

(3.13) $$\lim_{n\to\infty} \frac{S_n}{n} = \lim_{n\to\infty} \frac{1}{n}\sum_{k=0}^{n-1} \xi(X_k^t) = \int \xi(x)\pi(dx) \quad \text{a.s.}$$

Under either one of the statements (1) and (2) one must have
$$\int \xi(x)\pi(dx) = 0.$$

Hence (3.13) implies that
$$\lim_{n\to\infty} \max_{k\leq M\sqrt{n}} P\{|S_k| > \epsilon\sqrt{n}\} = 0.$$

Thus we have proved (3.12).

By Theorem 2.3, to prove the equivalence between (1) and (2), we need only to show that $\left\{\sum_{k=1}^n \xi(X_k)/\sqrt{n}\right\}_{n\geq 1}$ is bounded in probability iff so is $\left\{\sum_{k=1}^n \xi(X_k^t)/\sqrt{n}\right\}_{n\geq 1}$. By Lemma 3.3,

(3.14) $$\mathcal{L}\Big(\sum_{k=1}^n \xi(X_k^t)\Big) = \mathcal{L}\Big(\sum_{k=1}^{T_n} \delta_k \xi(X_k)\Big).$$

Take (3.12) and (3.14) into account. To prove the equivalence between (1) and (2), therefore, it is enough to show that

(3.15) $$\left\{\sum_{k=1}^{T_n}(\delta_k - (1-t))\xi(X_k)/\sqrt{n}\right\}_{n\geq 1} \quad \text{is bounded in probability.}$$

Take $s > 0$ and $\lambda > (1-t)^{-1}$. We have

$$P\left\{\left|\sum_{k=1}^{T_n}(\delta_k - (1-t))\xi(X_k)\right| > s\sqrt{n}, \ T_n \leq \lambda n\right\}$$

$$\leq P\left\{\max_{l \leq \lambda n}\left|\sum_{k=1}^{l}(\delta_k - (1-t))\xi(X_k)\right| > s\sqrt{n}\right\}.$$

Let P^δ and E^δ denote, respectively, the probability and expectation w.r.t. $\{\delta_k\}_{k\geq 1}$ and $\{T_n\}_{n\geq 0}$. Conditionally on $\{X_n\}_{n\geq 0}$ we have by the martingale inequality that

$$P^\delta\left\{\max_{l \leq \lambda n}\left|\sum_{k=1}^{l}(\delta_k - (1-t))\xi(X_k)\right| > s\sqrt{n}\right\}$$

$$\leq \frac{1}{s^2 n} E^\delta\left(\sum_{k=1}^{[\lambda n]}(\delta_k - (1-t))\xi(X_k)\right)^2 = \frac{1}{s^2 n} t(1-t) \sum_{k=1}^{[\lambda n]} \xi^2(X_k).$$

Therefore,

$$P\left\{\left|\sum_{k=1}^{T_n}(\delta_k - (1-t))\xi(X_k)\right| > s\sqrt{n}, \ T_n \leq \lambda n\right\}$$

$$\leq \frac{1}{s^2 n} t(1-t)[\lambda n] \cdot \int \xi^2(x)\pi(dx) \leq s^{-2}\lambda t(1-t) \cdot \int \xi^2(x)\pi(dx).$$

By the law of the large numbers for i.i.d. sequences,

$$P\{T_n \leq \lambda n\} \longrightarrow 1 \ \text{ as } \ n \to \infty.$$

Hence we have

$$\limsup_{n \to \infty} P\left\{\left|\sum_{k=1}^{T_n}(\delta_k - (1-t))\xi(X_k)\right| > s\sqrt{n}\right\} \leq s^{-2}\lambda t(1-t) \int \xi^2(x)\pi(dx).$$

Letting $\lambda \longrightarrow (1-t)^{-1}$ yields

(3.16)
$$\limsup_{n \to \infty} P\left\{\left|\sum_{k=1}^{T_n}(\delta_k - (1-t))\xi(X_k)\right| > s\sqrt{n}\right\}$$
$$\leq s^{-2} t \int \xi^2(x)\pi(dx) \ \ \forall s > 0.$$

In particular, we have (3.15).

We now prove (3.11). One can easily see that (3.12) and the statement (1) imply

$$\frac{1-t}{\sqrt{n}} \sum_{k=1}^{T_n} \xi(X_k) \longrightarrow N(0, (1-t)\sigma^2) \quad \text{in distribution.}$$

and that (3.14) and the statement (2) imply

$$\frac{1}{\sqrt{n}} \sum_{k=1}^{T_n} \delta_k \xi(X_k) \longrightarrow N(0, \sigma_t^2) \quad \text{in distribution.}$$

Hence it follows from (3.16) that

$$N(0, \sigma_t^2) \longrightarrow N(0, \sigma^2) \quad \text{in distribution as} \quad t \to 0^+.$$

Equivalently, we have (3.11). ∎

Lemma 3.5. *For every $0 < t < 1$ and every $0 < s < 1$,*

$$(3.17) \qquad \sum_{n=1}^{\infty} s^n P_t^n = s(1-t) \sum_{n=1}^{\infty} (t + s(1-t))^{n-1} P^n.$$

Proof. The proof is similar to that of Lemma 8.2.3 of Meyn-Tweedie (1993). By (3.7) we have

$$(3.18) \quad \sum_{n=1}^{\infty} s^n P_t^n = \sum_{n=1}^{\infty} s^n \sum_{k=1}^{\infty} P\{T_n = k\} P^k = \sum_{k=1}^{\infty} \Big(\sum_{n=1}^{\infty} s^n P\{T_n = k\} \Big) P^k.$$

Let

$$A(z) = \sum_{k=1}^{\infty} z^k P\{T_1 = k\} \quad \text{and} \quad B(z) = \sum_{k=1}^{\infty} z^k \sum_{n=1}^{\infty} s^n P\{T_n = k\} \qquad |z| < 1.$$

Then,
$$A(z) = \sum_{k=1}^{\infty} z^k (1-t) t^{k-1} = \frac{z(1-t)}{1-tz}$$

and therefore,
$$B(z) = \sum_{n=1}^{\infty} s^n \sum_{k=1}^{\infty} z^k P\{T_n = k\} = \sum_{n=1}^{\infty} s^n \big(A(z)\big)^n = \frac{sA(z)}{1-sA(z)}$$
$$= \frac{s(1-t)z}{1-(t+s(1-t))z} = s(1-t) \sum_{k=1}^{\infty} (t+s(1-t))^{k-1} z^k.$$

By the definition of $B(z)$ we must have
$$\sum_{n=1}^{\infty} s^n P\{T_n = k\} = s(1-t)(t+s(1-t))^{k-1} \quad \forall k \geq 1.$$

Hence, the lemma follows from (3.18). ∎

Proof of Theorem 3.1. Note that
$$\frac{1}{n} E_\pi S_n^2 = \int \xi^2(x) \pi(dx) + 2 \sum_{k=1}^{n-1} \left(1 - \frac{k}{n}\right) \int \xi(x) P^k \xi(x) \pi(dx) \quad \forall n \geq 1.$$

It follows from Kronecker's lemma that

(3.19) $$\lim_{n \to \infty} \frac{1}{n} E_\pi S_n^2 = \int \xi^2(x) \pi(dx) + 2 \sum_{k=1}^{\infty} \int \xi(x) P^k \xi(x) \pi(dx).$$

In particular, Chebyshev's inequality implies that $\{S_n/\sqrt{n}\}_{n \geq 1}$ is bounded in probability. By Theorem 2.3, (3.1) holds.

We proceed to prove (3.2) step by step.

Step 1. We first prove (3.2) under the additional assumption that $\{X_n\}_{n \geq 0}$ has an atom α such that

(3.20) $$\xi(x) \neq 0 \quad \forall x \in \alpha.$$

Consider (2.6). By the definition (I-3.6) of $i_\alpha(n)$ one can see that

(3.21) $$\{i_\alpha(n) \geq k\} = \{\tau_\alpha(k) \leq n-1\} \quad \forall k \geq 1.$$

This in particular implies that for each $n \geq 1$, $i_\alpha(n) + 1$ is a stopping time w.r.t. the i.i.d. random variables $\{\xi_k\}_{k\geq 1}$ given in (2.2). According to Theorem 2.1, $E\xi_1 = 0$ and $E\xi_1^2 < +\infty$. By Wald's equation (see, e.g. p.139, Theorem 3 in Chow-Teicher (1978)),

(3.22) $$E_\pi \left(\sum_{k=1}^{i_\alpha(n)+1} \xi_k \right)^2 = E_\pi(i_\alpha(n)+1) \cdot E\xi_1^2$$
$$= ((n-1)\pi(\alpha)+1)E\xi_1^2 = ((n-1)\pi(\alpha)+1)\pi(\alpha)^{-1}\sigma^2 \quad \forall n \geq 1,$$

where the last equality follows from (2.1) in Theorem 2.1. On the other hand, by a lemma on p.90 in Chung (1967),

$$\lim_{n\to\infty} \frac{1}{n} E_\pi \max_{k\leq n} \xi_k^2 = 0$$

which implies that

$$\lim_{n\to\infty} \frac{1}{n} E_\pi \xi_{i_\alpha(n)}^2 I_{\{i_\alpha(n)\geq 1\}} = \lim_{n\to\infty} \frac{1}{n} E_\pi \xi_{i_\alpha(n)+1}^2 = 0.$$

Hence, it follows from (3.22) that

(3.23) $$\lim_{n\to\infty} \frac{1}{n} E_\pi \left(\sum_{k=1}^{i_\alpha(n)-1} \xi_k \right)^2 = \sigma^2.$$

Note that under the condition (ii')

$$\sum_{n=1}^{\infty} \int \xi(x) P^n \xi(x) \pi(dx) = \int \sum_{n=1}^{\infty} \xi(x) P^n \xi(x) \pi(dx).$$

By (2.6), (3.19) and (3.23) we will have (3.2) if we prove that

(3.24) $$\lim_{n\to\infty} \frac{1}{n} E_\pi \left(\sum_{j=l_\alpha(n)+1}^{n-1} \xi(X_j) \right)^2 = 0$$

and

(3.25) $$\lim_{n\to\infty} \frac{1}{n} E_\pi S^2_{n\wedge(\tau_\alpha+1)} = 0.$$

We now prove (3.24). By the Markov property and the definition of $l_\alpha(n)$,

$$E_\pi\Big(\sum_{j=l_\alpha(n)+1}^{n-1} \xi(X_j)\Big)^2 = \sum_{k=2}^{n-1} E_\pi\Big(\sum_{j=l_\alpha(n)+1}^{n-1} \xi(X_j)\Big)^2 I_{\{l_\alpha(n)=n-k\}}$$

$$= \sum_{k=2}^{n-1} E_\pi\Big(\sum_{j=n-k+1}^{n-1} \xi(X_j)\Big)^2 I_{\{X_{n-k}\in\alpha, X_{n-k+1}\notin\alpha,\cdots,X_{n-1}\notin\alpha\}}$$

$$= \pi(\alpha) \sum_{k=2}^{n-1} E_\alpha\Big(\sum_{j=1}^{k-1} \xi(X_j)\Big)^2 I_{\{\tau_\alpha\geq k\}}$$

$$\leq \pi(\alpha) \sum_{k=1}^{n} E_\alpha\Big(\sum_{j=1}^{k} \xi(X_j)\Big)^2 I_{\{\tau_\alpha\geq k+1\}}.$$

The condition (i) implies that for every $N \geq 1$,

$$\lim_{n\to\infty} \frac{1}{n} \sum_{k=N+1}^{n} E_\alpha\Big(\sum_{j=1}^{N} \xi(X_j)\Big)^2 I_{\{\tau_\alpha\geq k+1\}} = 0.$$

To prove (3.24), therefore, it is sufficient to show that

(3.26) $$\limsup_{N\to\infty} \limsup_{n\to\infty} \frac{1}{n} \sum_{k=N+1}^{n} E_\alpha\Big(\sum_{j=N+1}^{k} \xi(X_j)\Big)^2 I_{\{\tau_\alpha\geq k+1\}} = 0.$$

For each $N+1 \leq k \leq n$,

$$E_\alpha\Big(\sum_{j=N+1}^{k} \xi(X_j)\Big)^2 I_{\{\tau_\alpha\geq k+1\}} \leq E_\alpha\Big(\sum_{j=N+1}^{k} \xi(X_j) I_{\{\tau_\alpha\geq j\}}\Big)^2$$

(3.27) $$\leq E_\alpha \sum_{j=N+1}^{\infty} \xi^2(X_j) I_{\{\tau_\alpha\geq j\}}$$

$$+ 2 E_\alpha\Big(\sum_{i=N+1}^{k} \sum_{j=i+1}^{k} \xi(X_i) I_{\{\tau_\alpha\geq i\}} \xi(X_j) I_{\{\tau_\alpha\geq j\}}\Big).$$

Given $i, j \geq 1$, on $\{\tau_\alpha \geq i+1\}$ we have

(3.28) $$\xi(X_{i+j})I_{\{\tau_\alpha \geq i+j\}} = \Big(\xi(X_j)I_{\{\tau_\alpha \geq j\}}\Big) \circ \theta^i,$$

where θ is the shift operator of $\{X_n\}_{n \geq 0}$. Therefore,

$$E_\alpha\Big(\sum_{i=N+1}^{k}\sum_{j=i+1}^{k} \xi(X_i)I_{\{\tau_\alpha \geq i\}}\xi(X_j)I_{\{\tau_\alpha \geq j\}}\Big)$$
$$= \sum_{i=N+1}^{k}\sum_{j=1}^{k-i} E_\alpha\Big(\xi(X_i)I_{\{\tau_\alpha \geq i+1\}}\xi(X_{i+j})I_{\{\tau_\alpha \geq i+j\}}\Big)$$
$$= \sum_{i=N+1}^{k}\sum_{j=1}^{k-i} E_\alpha\Big\{\xi(X_i)I_{\{\tau_\alpha \geq i+1\}}E_{X_i}\Big(\xi(X_j)I_{\{\tau_\alpha \geq j\}}\Big)\Big\}$$
$$= \sum_{i=N+1}^{k} E_\alpha\Big\{\xi(X_i)I_{\{\tau_\alpha \geq i+1\}}\sum_{j=1}^{k-i}E_{X_i}\Big(\xi(X_j)I_{\{\tau_\alpha \geq j\}}\Big)\Big\},$$

where the second equality follows from (3.28) and Markov property. By the first-entrance decomposition ((8.20) in Meyn-Tweedie (1993)),

$$E_x\Big(\xi(X_j)I_{\{\tau_\alpha \geq j\}}\Big) = P^j \xi(x) - \sum_{l=1}^{j-1} P_x\{\tau_\alpha = l\} P^{j-l}\xi(\alpha) \quad \forall x \in E.$$

Hence we obtain that

$$E_\alpha\Big(\sum_{i=N+1}^{k}\sum_{j=i+1}^{k} \xi(X_i)I_{\{\tau_\alpha \geq i\}}\xi(X_j)I_{\{\tau_\alpha \geq j\}}\Big)$$
$$= \sum_{i=N+1}^{k} E_\alpha\Big(\xi(X_i)I_{\{\tau_\alpha \geq i+1\}}\sum_{j=1}^{k-i}P^j\xi(X_i)\Big)$$
(3.29) $$\quad - \sum_{i=N+1}^{k} E_\alpha\Big(\xi(X_i)I_{\{\tau_\alpha \geq i+1\}}\sum_{j=1}^{k-i}\sum_{l=1}^{j-1}P_{X_i}\{\tau_\alpha = l\}P^{j-l}\xi(\alpha)\Big)$$
$$= \sum_{i=N+1}^{k} E_\alpha\Big(\xi(X_i)I_{\{\tau_\alpha \geq i+1\}}\sum_{j=1}^{k-i}P^j\xi(X_i)\Big)$$
$$\quad - \sum_{i=N+1}^{k} E_\alpha\Big(\xi(X_i)I_{\{\tau_\alpha \geq i+1\}}\sum_{l=1}^{k-i-1}P_{X_i}\{\tau_\alpha = l\}\sum_{j=1}^{k-i-l}P^j\xi(\alpha)\Big).$$

Note that (3.20) and the condition (ii') imply the convergence of the constant series $\sum_{j=1}^{\infty} P^j \xi(\alpha)$. In particular, there exists $D > 0$ such that

$$\sup_{k \geq 1} \left| \sum_{j=1}^{k} P^j \xi(\alpha) \right| \leq D. \tag{3.30}$$

From (3.29) we have

$$\left| E_\alpha \Big(\sum_{i=N+1}^{k} \sum_{j=i+1}^{k} \xi(X_i) I_{\{\tau_\alpha \geq i\}} \xi(X_j) I_{\{\tau_\alpha \geq j\}} \Big) \right|$$

$$\leq \sum_{i=N+1}^{k} E_\alpha \Big(\Big| \sum_{j=1}^{k-i} \xi(X_i) P^j \xi(X_i) \Big| I_{\{\tau_\alpha \geq i+1\}} \Big)$$

$$+ D \sum_{i=N+1}^{k} E_\alpha |\xi(X_i)| I_{\{\tau_\alpha \geq i+1\}} \tag{3.31}$$

$$\leq \sum_{i=N+1}^{k} a_{i,k-i} + D \sum_{i=N+1}^{\infty} E_\alpha |\xi(X_i)| I_{\{\tau_\alpha \geq i+1\}},$$

where we write

$$a_{i,k} \equiv E_\alpha \Big(\Big| \sum_{j=1}^{k} \xi(X_i) P^j \xi(X_i) \Big| I_{\{\tau_\alpha \geq i+1\}} \Big). \tag{3.32}$$

Combining (3.27) and (3.31) gives that for $n > N$,

$$\frac{1}{n} \sum_{k-N+1}^{n} E_\alpha \Big(\sum_{j-N+1}^{k} \xi(X_j) \Big)^2 I_{\{\tau_\alpha \geq k+1\}}$$

$$\leq \sum_{j=N+1}^{\infty} E_\alpha \xi^2(X_j) I_{\{\tau_\alpha \geq j\}} + 2D \sum_{i=N+1}^{\infty} E_\alpha |\xi(X_i)| I_{\{\tau_\alpha \geq i+1\}}$$

$$+ \frac{2}{n} \sum_{k=N+1}^{n} \sum_{i=N+1}^{k} a_{i,k-i}.$$

By Theorem 10.4.9 in Meyn-Tweedie (1993),

$$\sum_{j=1}^{\infty} E_\alpha \xi^2(X_j) I_{\{\tau_\alpha \geq j\}} = E_\alpha \sum_{j=1}^{\tau_\alpha} \xi^2(X_j) = \pi(\alpha)^{-1} \int \xi^2(x) \pi(dx) < +\infty$$

and similarly,

$$\sum_{i=1}^{\infty} E_\alpha |\xi(X_i)| I_{\{\tau_\alpha \geq i+1\}} \leq E_\alpha \sum_{i=1}^{\tau_\alpha} |\xi(X_i)| = \pi(\alpha)^{-1} \int |\xi(x)| \pi(dx) < +\infty.$$

Consequently,

$$\lim_{N \to \infty} \sum_{j=N+1}^{\infty} E_\alpha \xi^2(X_j) I_{\{\tau_\alpha \geq j\}} = \lim_{N \to \infty} \sum_{i=N+1}^{\infty} E_\alpha |\xi(X_i)| I_{\{\tau_\alpha \geq i+1\}} = 0.$$

To prove (3.26), therefore, it is enough to show that

(3.33) $$\limsup_{N \to \infty} \limsup_{n \to \infty} \frac{1}{n} \sum_{k=N+1}^{n} \sum_{i=N+1}^{k} a_{i,k-i} = 0.$$

Note that

$$\sum_{k=N+1}^{n} \sum_{i=N+1}^{k} a_{i,k-i} = \sum_{k=1}^{n-N-1} \sum_{i=N+1}^{n-k} a_{i,k}$$

$$\leq \sum_{k=1}^{n-N-1} \sum_{i=N+1}^{\infty} a_{i,k} \leq n \sup_{k \geq 1} \sum_{i=N+1}^{\infty} a_{i,k}.$$

Hence, in order that (3.33) hold, we need only to prove

(3.34) $$\lim_{N \to \infty} \sup_{k \geq 1} \sum_{i=N+1}^{\infty} a_{i,k} = 0.$$

By condition (ii'), $\left\{ \sum_{j=1}^{k} \xi(x) P^j \xi(x) \right\}_{k \geq 1}$ is uniformly integrable. That is, for any given $\epsilon > 0$ there exists $M > 0$ such that

(3.35) $$\sup_{k \geq 1} \int_{\{|\sum_{j=1}^{k} \xi(x) P^j \xi(x)| > M\}} \left| \sum_{j=1}^{k} \xi(x) P^j \xi(x) \right| \pi(dx) < \epsilon.$$

From (3.32), we can get that

$$a_{i,k} \leq E_\alpha \left(\left| \sum_{j=1}^{k} \xi(X_i) P^j \xi(X_i) \right| I_{\{|\sum_{j=1}^{k} \xi(X_i) P^j \xi(X_i)| > M\}} I_{\{\tau_\alpha \geq i\}} \right)$$
$$+ M P_\alpha \{\tau_\alpha \geq i\}.$$

Applying Theorem 10.4.9 in Meyn-Tweedie (1993) again yields

$$\sum_{i=N+1}^{\infty} a_{i,k}$$

$$\leq \pi(\alpha)^{-1} \int_{\{|\sum_{j=1}^{k} \xi(x) P^j \xi(x)| > M\}} \left| \sum_{j=1}^{k} \xi(x) P^j \xi(x) \right| \pi(dx)$$

$$+ M \sum_{i=N+1}^{\infty} P_\alpha\{\tau_\alpha \geq i\}$$

$$\leq \pi(\alpha)^{-1} \epsilon + M \sum_{i=N+1}^{\infty} P_\alpha\{\tau_\alpha \geq i\}.$$

Note that the positivity of $\{X_n\}_{n \geq 0}$ implies that (Theorem 10.4.9 in Meyn-Tweedie (1993))

$$\sum_{i=1}^{\infty} P_\alpha\{\tau_\alpha \geq i\} = E_\alpha \tau_\alpha = \frac{1}{\pi(\alpha)} < +\infty.$$

Hence

$$\lim_{N \to \infty} \sum_{i=N+1}^{\infty} P_\alpha\{\tau_\alpha \geq i\} = 0.$$

Therefore, (3.34) holds.

We now come to the proof of (3.25). Consider

(3.36)
$$E_\pi S_{n \wedge (\tau_\alpha + 1)}^2 = \sum_{k=0}^{n-1} E_\pi \xi^2(X_k) I_{\{\tau_\alpha \geq k\}}$$
$$+ 2 E_\pi \left(\sum_{j=0}^{n-1} \sum_{k=j+1}^{n-1} \xi(X_j) I_{\{\tau_\alpha \geq j\}} \xi(X_k) I_{\{\tau_\alpha \geq k\}} \right).$$

The estimate

$$E_\pi \xi^2(X_n) I_{\{\tau_\alpha \geq n\}} \leq L^2 P_\pi\{\tau_\alpha \geq n\} + \int_{|\xi(x)| > L} \xi^2(x) \pi(dx) \quad (L > 0)$$

and the condition (i) imply that

$$\lim_{n\to\infty} E_\pi \xi^2(X_n) I_{\{\tau_\alpha \geq n\}} = 0.$$

Hence

(3.37) $$\lim_{n\to\infty} \frac{1}{n} \sum_{k=0}^{n-1} E_\pi \xi^2(X_k) I_{\{\tau_\alpha \geq k\}} = 0.$$

A computation similar to that for (3.29) gives

$$E_\pi \Big(\sum_{j=0}^{n-1} \sum_{k=j+1}^{n-1} \xi(X_j) I_{\{\tau_\alpha \geq j\}} \xi(X_k) I_{\{\tau_\alpha \geq k\}} \Big)$$

$$= \sum_{j=0}^{n-1} E_\pi \Big(\sum_{k=1}^{n-j-1} \xi(X_j) P^k \xi(X_j) I_{\{\tau_\alpha \geq j+1\}} \Big)$$

$$- \sum_{j=0}^{n-1} E_\pi \Big(\xi(X_j) I_{\{\tau_\alpha \geq j+1\}} \sum_{l=1}^{n-j-2} P_{X_j}\{\tau_\alpha = l\} \sum_{k=1}^{n-j-l-1} P^k \xi(\alpha) \Big).$$

In view of (3.30) and (3.35) we have

$$\frac{1}{n} \Big| E_\pi \Big(\sum_{j=0}^{n-1} \sum_{k=j+1}^{n-1} \xi(X_j) I_{\{\tau_\alpha \geq j\}} \xi(X_k) I_{\{\tau_\alpha \geq k\}} \Big) \Big|$$

$$\leq \frac{1}{n} \sum_{j=0}^{n-1} E_\pi \Big(\Big| \sum_{k=1}^{n-j-1} \xi(X_j) P^k \xi(X_j) \Big| I_{\{\tau_\alpha \geq j+1\}} \Big)$$

$$+ \frac{D}{n} \sum_{j=0}^{n-1} E_\pi |\xi(X_j)| I_{\{\tau_\alpha \geq j+1\}}$$

$$\leq \epsilon + \frac{M}{n} \sum_{j=0}^{n-1} P_\pi\{\tau_\alpha \geq j+1\} + \frac{D}{n} \sum_{j=0}^{n-1} E_\pi |\xi(X_j)| I_{\{\tau_\alpha \geq j+1\}}.$$

One can see now that the second and the third terms on the right-hand side tend to zero as $n \longrightarrow \infty$. Therefore we have proved that

(3.38) $$\lim_{n\to\infty} \frac{1}{n} E_\pi \Big(\sum_{j=0}^{n-1} \sum_{k=j+1}^{n-1} \xi(X_j) I_{\{\tau_\alpha \geq j\}} \xi(X_k) I_{\{\tau_\alpha \geq k\}} \Big) = 0.$$

Hence (3.25) follows from (3.36), (3.37) and (3.38).

Step 2. We assume additionally that there exists $C \in \mathcal{E}^+$ such that

$$(3.39) \qquad P \geq bI_C \otimes \pi_C$$

for some $0 < b < 1$ and that

$$(3.40) \qquad \delta \equiv \inf_{x \in C} |\xi(x)| > 0,$$

where the probability measure π_C is given in (I-1.5). According to Section I-2, the minorization (3.39) generates a split chain (in the classic sense) on $E \times I$, which is ergodic (Theorem I-2.5) and has an atom $\alpha^* = C \times \{1\}$. When viewed as a function on $E \times I$, ξ satisfies (3.20) with α^* instead of α. Since

$$(3.41) \qquad \int \xi^2(x) \pi^*(d(x,y)) = \int \xi^2(x) \pi(dx) < +\infty,$$

this split chain satisfies the condition (i). We now show that it also satisfies (ii'). A computation similar to (A.1) in Appendix (with $m = 1$ and with ξ instead of h) gives

$$(3.42) \qquad \begin{aligned} \tilde{P}^n \xi(x,y) &= I_0(y) \int Q(x, dx_1) P^{n-1} \xi(x_1) \\ &+ I_1(y) \int \pi_C(dx_1) P^{n-1} \xi(x_1) \end{aligned}$$

for all $(x,y) \in E \times I$ and $n \geq 2$, where Q is, according to the construction in Section I-2, a transition probability on $(E, \mathcal{E})$ satisfying

$$(3.43) \quad P(x, A) = (1 - bI_C(x))Q(x, A) + bI_C(x)\pi_C(A) \quad x \in E, \ A \in \mathcal{E}.$$

Given $\epsilon > 0$ by the condition (ii') there exists $N \geq 1$ such that

$$(3.44) \qquad \int \left| \sum_{j=k+1}^{k+n} \xi(x) P^j \xi(x) \right| \pi(dx) < \epsilon$$

holds for all $k \geq N$ and all $n \geq 1$.

Combining (3.42) and (3.43) gives

$$\int \Big| \sum_{j=k+1}^{k+n} \xi(x) \tilde{P}^j \xi(x,y) \Big| \pi^*(d(x,y))$$

$$= \int_{\{y=0\}} \pi^*(d(x,y)) \Big| \int Q(x, dx_1) \sum_{j=k}^{k+n-1} \xi(x) P^j \xi(x_1) \Big|$$

$$+ \int_{\{y=1\}} \pi^*(d(x,y)) \Big| \int \pi_C(dx_1) \sum_{j=k}^{k+n-1} \xi(x) P^j \xi(x_1) \Big|$$

$$= \int \pi(dx) \Big| (1 - bI_C(x)) \int Q(x, dx_1) \sum_{j=k}^{k+n-1} \xi(x) P^j \xi(x_1) \Big|$$

$$+ \int \pi(dx) \Big| bI_C(x) \int \pi_C(dx_1) \sum_{j=k}^{k+n-1} \xi(x) P^j \xi(x_1) \Big|$$

$$\leq \int \pi(dx) \Big| \Big((1 - bI_C(x)) \int Q(x, dx_1)$$

$$+ bI_C(x) \int \pi_C(dx_1) \Big) \sum_{j=k}^{k+n-1} \xi(x) P^j \xi(x_1) \Big|$$

$$+ 2 \int \pi(dx) \Big| bI_C(x) \int \pi_C(dx_1) \sum_{j=k}^{k+n-1} \xi(x) P^j \xi(x_1) \Big|$$

$$\leq \int \pi(dx) \Big| \int P(x, dx_1) \sum_{j=k}^{k+n-1} \xi(x) P^j \xi(x_1) \Big|$$

$$+ 2b \Big(\int |\xi(x)| \pi(dx) \Big) \Big(\int \Big| \sum_{j=k}^{k+n-1} P^j \xi(x) \Big| \pi_C(dx) \Big).$$

From (3.40) and (3.44),

$$(3.45) \qquad \int \Big| \sum_{j=k+1}^{k+n} P^j \xi(x) \Big| \pi_C(dx) < \delta^{-1} \pi(C)^{-1} \epsilon$$

for all $k \geq N$ and all $n \geq 1$. Therefore, by (3.44) and (3.45),

$$\int \Big| \sum_{j=k+1}^{k+n} \xi(x) \tilde{P}^j \xi(x,y) \Big| \pi^*(d(x,y))$$

$$\leq \int \Big| \sum_{j=k+1}^{k+n} \xi(x) P^j \xi(x) \Big| \pi(dx) + 2b\delta^{-1}\pi(C)^{-1}\epsilon \int |\xi(x)| \pi(dx)$$

$$\leq \epsilon + 2b\delta^{-1}\pi(C)^{-1}\epsilon \int |\xi(x)| \pi(dx)$$

for all $k \geq N+1$ and all $n \geq 1$. Since $\epsilon > 0$ is arbitrary, we must have that

$$\sum_{n=1}^{\infty} \xi(\cdot) \tilde{P}^n \xi(\cdot,\cdot) \quad \text{converges in} \quad L^1(\pi^*).$$

By the conclusion achieved at Step 1 and (3.41) we have

$$\sigma^2 = \int \xi^2(x) \pi^*(d(x,y)) + 2 \int \sum_{n=1}^{\infty} \xi(x) \tilde{P}^n \xi(x,y) \pi^*(d(x,y))$$

$$= \int \xi^2(x) \pi(dx) + 2 \sum_{n=1}^{\infty} \int \xi(x) \tilde{P}^n \xi(x,y) \pi^*(d(x,y)).$$

By (3.42) and (3.43), for each $n \geq 1$,

$$\int \xi(x) \tilde{P}^n \xi(x,y) \pi^*(d(x,y))$$

$$= \int_{\{y=0\}} \pi^*(d(x,y)) \xi(x) \int Q(x,dx_1) P^{n-1}\xi(x_1)$$

$$+ \int_{\{y=1\}} \pi^*(d(x,y)) \xi(x) \int \pi_C(dx_1) P^{n-1}\xi(x_1)$$

$$= \int \pi(dx)(1 - bI_C(x))\xi(x) \int Q(x,dx_1) P^{n-1}\xi(x_1) \cdot$$

$$+ \int \pi(dx) b I_C(x) \xi(x) \int \pi_C(dx_1) P^{n-1}\xi(x_1)$$

$$= \int \pi(dx) \xi(x) \int P(x,dx_1) P^{n-1}\xi(x_1)$$

$$= \int \xi(x) P^n \xi(x) \pi(dx).$$

Hence (3.2) holds.

Step 3. We now prove (3.2) in the general context. Without loss of generality we may assume that $\{x;\ \xi(x) \neq 0\} \in \mathcal{E}^+$. By Theorem 5.2.1 in Meyn-Tweedie (1993), there exists a sub-π small set C of $\{X_n\}_{n\geq 0}$ such that (3.40) holds. For each $0 < t < 1$, let P_t be the resolvent kernel of P given in (3.6) and let $\{X_n^t\}_{n\geq 0}$ be a resolvent chain conrresponding to P_t. Then for each $0 < t < 1$, P_t satisfies the additional assumption at Step 2 with every sub-π small set C of $\{X_n\}_{n\geq 0}$ satisfying (3.40). By Lemma 3.2 $\{X_n^t\}_{n\geq 1}$ is ergodic with the same invariant distribution π. Consequently P_t satisfies the condition (i). We now show that it also satisfies the condition (ii'). That is, for each $0 < t < 1$,

$$(3.46) \qquad \sum_{n=1}^{\infty} \xi(\cdot) P_t^n \xi(\cdot) \quad \text{converges in} \quad L^1(\pi).$$

To do this, it is enough to prove that given $\epsilon > 0$, there exists $N \geq 1$ such that

$$(3.47) \qquad \int \left| \sum_{n=N+1}^{\infty} s^n \xi(x) P_t^n \xi(x) \right| \pi(dx) \leq \epsilon$$

holds for all $0 < s < 1$.

By the condition (ii') we can choose $N \geq 1$ such that (3.44) holds for all $k \geq N$ and all $n \geq 1$. Applying Lemma 3.5 and taking (3.7) and (3.8) into account we have

$$\sum_{n=N+1}^{\infty} s^n P_t^n = s^N P_t^N \sum_{n=1}^{\infty} s^n P_t^n$$

$$= (1-t) s^{N+1} P_t^N \sum_{n=1}^{\infty} (t + s(1-t))^{n-1} P^n$$

$$= (1-t) s^{N+1} \sum_{k=N}^{\infty} P\{T_N = k\} \sum_{n=1}^{\infty} (t + s(1-t))^{n-1} P^{n+k}.$$

Therefore,

$$
\begin{aligned}
(3.48)\quad & \int \Big| \sum_{n=N+1}^{\infty} s^n \xi(x) P_t^n \xi(x) \Big| \pi(dx) \\
& \leq \sum_{k=N}^{\infty} P\{T_N = k\} \int \Big| \sum_{n=1}^{\infty} (t+s(1-t))^{n-1} \xi(x) P^{n+k} \xi(x) \Big| \pi(dx).
\end{aligned}
$$

Write $z = t + s(1-t)$. For every $k \geq N$ and $p > 1$,

$$
\sum_{n=1}^{p} z^{n-1} P^{n+k} \xi(x) = (1-z) \sum_{n=1}^{p} z^{n-1} \sum_{j=k+1}^{n+k} P^j \xi(x) + z^p \sum_{j=k+1}^{k+p} P^j \xi(x).
$$

By (3.44) we have

$$
\int \Big| \sum_{n=1}^{p} z^{n-1} \xi(x) P^{n+k} \xi(x) \Big| \pi(dx)
$$

$$
\leq (1-z) \sum_{n=1}^{p} z^{n-1} \int \Big| \sum_{j=k+1}^{n+k} \xi(x) P^j \xi(x) \Big| \pi(dx)
$$

$$
+ z^p \int \Big| \sum_{j=k+1}^{k+p} \xi(x) P^j \xi(x) \Big| \pi(dx)
$$

$$
\leq \epsilon(1-z) \sum_{n=1}^{p} z^{n-1} + z^p \epsilon \leq \epsilon + z^p \epsilon.
$$

Letting $p \longrightarrow +\infty$ gives

$$
\int \Big| \sum_{n=1}^{\infty} z^{n-1} \xi(x) P^{n+k} \xi(x) \Big| \pi(dx) \leq \epsilon \quad \forall k \geq N.
$$

This, along with (3.48), implies (3.47).

Applying the conclusion achieved at Step 2 to the chain $\{X_n^t\}_{n \geq 1}$ we have that for each $0 < t < 1$,

$$
\sum_{k=0}^{n-1} \xi(X_k^t)/\sqrt{n} \longrightarrow N(0, \sigma_t^2) \quad \text{in distribution}
$$

and
$$\sigma_t^2 = \int \xi^2(x)\pi(dx) + 2\int \sum_{n=1}^{\infty} \xi(x)P_t^n\xi(x)\pi(dx).$$

As a consequence of Lemma 3.5, for each $0 < s < 1$ we have

$$\int \sum_{n=1}^{\infty} s^n \xi(x)P_t^n\xi(x)\pi(dx) = s(1-t)\int \sum_{n=1}^{\infty}(t+s(1-t))^{n-1}\xi(x)P^n\xi(x)\pi(dx).$$

Letting $s \longrightarrow 1$ yields

$$\int \sum_{n=1}^{\infty} \xi(x)P_t^n\xi(x)\pi(dx) = (1-t)\int \sum_{n=1}^{\infty} \xi(x)P^n\xi(x)\pi(dx).$$

Therefore

$$\sigma_t^2 = \int \xi^2(x)\pi(dx) + 2(1-t)\int \sum_{n=1}^{\infty} \xi(x)P^n\xi(x)\pi(dx) \quad \forall 0 < t < 1.$$

Consequently,

$$\lim_{t \to 0^+} \sigma_t^2 = \int \xi^2(x)\pi(dx) + 2\int \sum_{n=1}^{\infty} \xi(x)P^n\xi(x)\pi(dx).$$

Hence (3.2) follows from (3.11) in Lemma 3.4. ∎

II-4. The CLT and the ergodicity conditions

Intuitively, the stronger the ergodicity assumption on $\{X_n\}_{n\geq 1}$, the more likely the CLT holds for the same function $\xi\colon E \longrightarrow \mathbf{R}$. Recently, de Acosta ([4]) pointed out that under ergodicity of degree 2, the CLT holds for every bounded and mean zero function ξ. Our first result in this section claims the ergodicity of degree 2 is the weakest one satisfying this requirement. More precisely, we have the following:

Theorem 4.1. *For an ergodic Markov chain $\{X_n\}_{n\geq 0}$, the following three statements are equivalent:*

(i) $E_\pi \tau_A < +\infty$ *for every* $A \in \mathcal{E}^+$;

(ii) $\sum_{n=1}^\infty \int ||P^n(x,\cdot) - \pi||_V \pi(dx) < +\infty$;

(iii) for every bounded function $\xi: E \longrightarrow \mathbf{R}$ *with*

(4.1) $$\int \xi(x)\pi(dx) = 0,$$

there exists $\sigma^2 \geq 0$ *such that*

$$S_n/\sqrt{n} \longrightarrow N(0, \sigma^2) \quad \text{in distribution.}$$

Further,

(4.2) $$\sigma^2 = \int \xi^2(x)\pi(dx) + 2\int \sum_{n=1}^\infty \xi(x)P^n\xi(x)\pi(dx)$$

whenever one of these statements holds.

Remark 4.2. Recall that the chain $\{X_n\}_{n\geq 0}$ is called ergodic of degree 2 if it is ergodic and the statement (i) holds. The meaning of Theorem 4.1 is twofold: one is that ergodicity of degree 2 is the weakest ergodicity which allows the CLT holds for every bounded and mean zero function ξ; another is that ergodicity of degree 2 can be defined otherwise. Namely, the chain $\{X_n\}_{n\geq 0}$ is ergodic of degree 2 if and only if it is ergodic and satisfies the statement (ii). In view of the way we define the other ergodicity conditions, this new definition seems to be more natural than the previous one.

Proof of Theorem 4.1. The direction "(i) $\Longrightarrow$ (ii)" follows from Proposition 2.1 in de Acosta ([4]). Notice that the statement (ii) implies the condition (ii') in Theorem 3.1 for every bounded ξ with (4.1). Therefore the direction "(ii) $\Longrightarrow$ (iii)" follows from Theorem 3.1 and, (4.2) holds for every bounded and mean zero ξ under the statement (ii). We need only to prove the direction "(iii) $\Longrightarrow$ (i)". Since for every $A \in \mathcal{E}^+$, by Theorem 5.2.2 in Meyn-Tweedie (1993) there is a small set $C \subset A$, it is enough to show that

$$\text{(4.3)} \qquad E_\pi \tau_C < +\infty$$

holds for every small set C.

First we show that (4.3) holds for every small set C of order 1. We can choose $0 < b < 1/2$ such that

$$\text{(4.4)} \qquad P \geq 2b I_C \otimes \nu$$

holds for some probability measure ν on $(E, \mathcal{E})$. In particular,

$$\text{(4.5)} \qquad P \geq b I_C \otimes \nu.$$

Consider the split chain (in the classical sense) $\{(X_n, Y_n)\}_{n \geq 0}$ generated by the minorization (4.5). From (I-2.14), the transition $\tilde{P}$ of $\{(X_n, Y_n)\}_{n \geq 0}$ is defined by

$$\tilde{P}((x,y), \bar{A}) = I_0(y) Q^*(x, \bar{A}) + I_1(y) \nu^*(\bar{A}) \qquad (x,y) \in E \times I, \quad \bar{A} \in \mathcal{E} \otimes \mathcal{I},$$

where

$$Q(x, A) = (1 - b I_C(x))^{-1}(P(x, A) - b I_C(x) \nu(A)) \qquad x \in E, \quad A \in \mathcal{E}.$$

From (4.4),

$$Q(x, \cdot) \geq \frac{b}{1-b} \nu \qquad \forall x \in C$$

which implies

$$Q^*(x, \cdot) \geq \frac{b}{1-b} \nu^* \qquad \forall x \in C.$$

Consequently,

(4.6) $$\tilde{P}((x,y),\cdot) \geq \frac{b}{1-b}\nu^* \quad \forall (x,y) \in C \times I.$$

Define $\xi\colon E \longrightarrow \mathbf{R}$ by

$$\xi(x) = \begin{cases} 1 & x \notin C \\ 1 - \frac{1}{\pi(C)} & x \in C. \end{cases}$$

Clearly, ξ is bounded and satisfies (4.1). Hence the CLT holds. By Theorem 2.2,

(4.7) $$\tilde{E}_{\alpha^*}\left(\sum_{k=1}^{\tau_{\alpha^*}} \xi(X_k)\right)^2 = \tilde{E}_{\nu^*}\left(\sum_{k=0}^{\tau(0)} \xi(X_k)\right)^2 < +\infty,$$

where $\alpha^* = C \times \{1\}$ and

$$\tau_{\alpha^*} = \inf\{n \geq 1; (X_n, Y_n) \in \alpha^*\}.$$

Define the stopping times $\{\tau_C(k)\}_{k \geq 0}$ as follows:

$$\begin{cases} \tau_C(0) = 0, \quad \tau_C(1) = \tau_C, \\ \tau_C(k+1) = \inf\{n > \tau_C(k); X_n \in C\} \quad (k \geq 1), \end{cases}$$

and define

$$T = \inf\{k \geq 1; (X_{\tau_C(k)}, Y_{\tau_C(k)}) \in \alpha^*\}.$$

Then,

(4.8) $$\sum_{k=1}^{\tau_{\alpha^*}} \xi(X_k) = \sum_{k=0}^{T-1} \sum_{j=\tau_C(k)+1}^{\tau_C(k+1)} \xi(X_j)$$
$$= \sum_{k=0}^{T-1}\left(\tau_C(k+1) - \tau_C(k) - \frac{1}{\pi(C)}\right) = \tau_{\alpha^*} - \frac{1}{\pi(C)}T \quad \text{a.s.}$$

Notice that under the law $\tilde{P}$, $\{(X_{\tau_C(k)}, Y_{\tau_C(k)})\}_{k\geq 0}$ becomes a Markov chain with the state space $C \times I$ and the transition probability determined by

$$\tilde{P}_{(x,y)}\{(X_{\tau_C}, Y_{\tau_C}) \in \cdot\} \quad (x,y) \in C \times I$$

and that α^* is an atom of this chain. From (4.6) we have

$$\tilde{P}_{(x,y)}\{(X_{\tau_C}, Y_{\tau_C}) \in \cdot\}$$
$$= \int \tilde{P}((x,y), d(x_1, y_1)) \tilde{P}_{(x_1,y_1)}\{(X_{\tau_C'}, Y_{\tau_C'}) \in \cdot\}$$
$$\geq \frac{b}{1-b} \tilde{P}_{\nu^*}\{(X_{\tau_C'}, Y_{\tau_C'}) \in \cdot\} \quad (x,y) \in C \times I,$$

where

$$\tau_C' = \inf\{n \geq 0; X_n \in C\}.$$

It means that $C \times I$ is a small set of the Markov chain $\{(X_{\tau_C(k)}, Y_{\tau_C(k)})\}_{k\geq 0}$. By Theorem 16.0.2 in Meyn-Tweedie (1993), the chain $\{(X_{\tau_C(k)}, Y_{\tau_C(k)})\}_{k\geq 0}$ is uniformly ergodic. In particular,

$$\tilde{E}_{\alpha^*} T^2 < +\infty.$$

This, along with (4.7) and (4.8), implies that

$$\tilde{E}_{\alpha^*} \tau_{\alpha^*}^2 < +\infty.$$

By Proposition 5.15 in Nummelin (1984) we have

$$\tilde{E}_{\pi^*} \tau_{\alpha^*} < +\infty.$$

Note that $\tau_C \leq \tau_{\alpha^*}$. Using (I-2.8) gives

$$E_\pi \tau_C = \tilde{E}_{\pi^*} \tau_C \leq \tilde{E}_{\pi^*} \tau_{\alpha^*} < +\infty.$$

We now prove (4.3) for the general small set C. Fix $0 < t < 1$, let the random variables $\{T_n\}_{n\geq 0}$ be given in (3.4), let P_t be the resolvent kernel given in (3.6) and let $\{X_n^t\}_{n\geq 0}$ be a resolvent chain corresponding to P_t. For the chain $\{X_n^t\}_{n\geq 0}$, C is a small set of order 1. According to Lemma 3.2, $\{X_n^t\}_{n\geq 0}$ is ergodic with the same invariant distribution π. By Lemma 3.4, the chain $\{X_n^t\}_{n\geq 0}$ satisfies the statement (iii). Hence,

$$E_\pi^t \tau_C^t < +\infty,$$

where E_π^t denotes the Markovian expectation generated by the transition P_t and the initial distribution π, and

$$\tau_C^t = \inf\{n \geq 1;\ X_n^t \in C\}.$$

According to (5.9) in de Acosta (1988),

$$\mathcal{L}(\{X_n^t\}_{n\geq 0}) = \mathcal{L}(\{X_{T_n}\}_{n\geq 0}).$$

Equivalently we have

(4.9) $$E\tau' < +\infty,$$

where $\tau' \equiv \inf\{n \geq 1;\ X_{T_n} \in C\}$ and E denotes the expectation w.r.t. both the random variables $\{T_n\}_{n\geq 0}$ and the Markov chain $\{X_n\}_{n\geq 0}$ (with the initial distribution π). Notice that $\{T_n\}_{n\geq 0}$ and $\{X_n\}_{n\geq 0}$ are independent. Therefore we can apply Wald's equation conditionally on $\{X_n\}_{n\geq 0}$. More precisely, if we view $\{X_n\}_{n\geq 0}$ for a moment as fixed (non-random) elements of E, then by definition for each $k \geq 1$, the event $\{\tau' = k\}$ depends only on $\{T_1, \cdots, T_k\}$. Applying Wald's equation (see, e.g. p.137, Theorem 1 in Chow-Teicher (1978)) to the i.i.d. sequence $\{T_k - T_{k-1}\}_{k\geq 1}$ and its stopping time τ' gives

$$E^\delta T_{\tau'} = ET_1 \cdot E^\delta \tau' = (1-t)^{-1} E^\delta \tau',$$

where E^δ denotes the expectation w.r.t. the random variables $\{T_n\}_{n\geq 0}$. Let $\{X_n\}_{n\geq 0}$ be a Markov chain again and take the independence between $\{T_n\}_{n\geq 0}$ and $\{X_n\}_{n\geq 0}$ into account. Applying Fubini's theorem to the above equality gives

$$ET_{\tau'} = (1-t)^{-1}E\tau' < +\infty.$$

On the other hand, by definition we have

$$T_{\tau'} = \inf\{T_n;\ n \geq 1\ \text{and}\ X_{T_n} \in C\} \geq \inf\{n \geq 1;\ X_n \in C\} = \tau_C.$$

Hence,

$$E_\pi \tau_C = E\tau_C \leq ET_{\tau'} < +\infty.$$

∎

We end this chapter by giving the following result which tells us that as far as the CLT is concerned, functionals of uniformly ergodic Markov chains behave like i.i.d. sequences.

Theorem 4.3. *Let $\{X_n\}_{n\geq 0}$ be an uniformly ergodic Markov chain and assume that the measurable map $\xi: E \longrightarrow \mathbf{R}$ satisfies*

(4.10) $$\int \xi(x)\pi(dx) = 0 \quad \text{and} \quad \int \xi^2(x)\pi(dx) < +\infty.$$

Then,

$$S_n/\sqrt{n} \longrightarrow N(0,\sigma^2) \quad \text{in distribution},$$

where

$$\sigma^2 = \int \xi^2(x)\pi(dx) + 2\int \sum_{n=1}^{\infty} \xi(x)P^n\xi(x)\pi(dx).$$

Proof. It is enough to verify that condition (ii') in Theorem 3.1 is satisfied. Consider the following integrability condition:

$$\int_C \pi(dx) E_x \Big(\sum_{k=0}^{\tau_C - 1} |\xi(X_k)| \Big)^2 < +\infty, \tag{4.11}$$

where $C \in \mathcal{E}^+$. Assumption (4.10) implies (4.11) for $C = E$. By Theorem 16.02 in Meyn-Tweedie (1993), E is small under the uniform ergodicity condition. Applying Theorem I-4.1 with $B = \mathbf{R}$, $\varphi(s) = s^2$ and $|\xi|$ instead of ξ we can see that (4.11) holds for all small sets C.

On the other hand, for fixed small set C,

$$\Big(\sum_{k=0}^{\tau_C - 1} |\xi(X_k)| \Big)^2 \geq \sum_{j=0}^{\tau_C - 1} \sum_{k=j}^{\tau_C - 1} |\xi(X_j)| |\xi(X_k)|$$

$$= \sum_{k=j}^{\infty} |\xi(X_j)| I_{\{\tau_C \geq j+1\}} \sum_{k=j}^{\tau_C - 1} |\xi(X_k)| \quad \text{a.s.}$$

Note that on the event $\{\tau_C \geq j + 1\}$,

$$\sum_{k=j}^{\tau_C - 1} |\xi(X_k)| = \Big(\sum_{k=0}^{\tau_C - 1} |\xi(X_k)| \Big) \circ \theta^j \quad \text{a.s.}$$

By the Markov property,

$$\int_C \pi(dx) E_x \Big(\sum_{k=0}^{\tau_C - 1} |\xi(X_k)| \Big)^2$$

$$\geq \int_C \pi(dx) \sum_{k=j}^{\infty} E_x \Big(|\xi(X_j)| I_{\{\tau_C \geq j+1\}} \sum_{k=j}^{\tau_C - 1} |\xi(X_k)| \Big)$$

$$= \int_C \pi(dx) \sum_{k=j}^{\infty} E_x \Big(|\xi(X_j)| I_{\{\tau_C \geq j+1\}} E_{X_j} \Big(\sum_{k=0}^{\tau_C - 1} |\xi(X_k)| \Big) \Big)$$

$$= \int_C \pi(dx) E_x \sum_{j=0}^{\tau_C - 1} |\xi(X_j)| E_{X_j} \Big(\sum_{k=0}^{\tau_C - 1} |\xi(X_k)| \Big)$$

$$= \int \pi(dx) |\xi(x)| E_x \sum_{k=0}^{\tau_C - 1} |\xi(X_k)|.$$

where the last step follows from Theorem 10.4.9 in Meyn-Tweedie (1993). We thus have

$$\int \pi(dx)|\xi(x)|E_x \sum_{k=0}^{\tau_C-1} |\xi(X_k)| < +\infty$$

for all small sets C. For every $A \in \mathcal{E}^+$, according to Theorem 5.2.2 in Meyn-Tweedie (1993), there is a small set C such that $C \subset A$. Therefore,

$$\int \pi(dx)|\xi(x)|E_x \sum_{k=0}^{\tau_A-1} |\xi(X_k)| < +\infty$$

for every $A \in \mathcal{E}^+$. Since uniform ergodicity implies ergodicity of degree 2,

$$\int \pi(dx) E_x \tau_A < +\infty \quad \forall A \in \mathcal{E}^+.$$

By Theorem 14.3.4 in Meyn-Tweedie (1993),

$$\sum_{n=1}^{\infty} \int \pi(dx)\pi(dy)|\xi(x)||P^n\xi(x) - P^n\xi(y)| < +\infty.$$

In particular,

$$\sum_{n=1}^{\infty} \int |\xi(x)||P^n\xi(x)|\pi(dx) < +\infty,$$

which implies the condition (ii') in Theorem 3.1. Hence, the desired conclusion follows from Theorem 3.1. ∎

Chapter III. The Law of the Iterated Logarithm

III-1. Introduction

In this chapter, $\{X_n\}_{n\geq 0}$ is always an ergodic Markov chain and ξ is a measurable map from E to $\mathbf{R}$ or more generally, to a separable Banach space B. As before,

$$S_n = \sum_{k=0}^{n-1} \xi(X_k), \quad n = 1, 2, \cdots.$$

The main purpose of this chapter is to study the strong limiting behavior of the sequence:

(1.1) $$\{S_n/\sqrt{2nL_2n}\}_{n\geq 1},$$

where the function $L_2 x$ $(x \geq 0)$ is defined by

$$L_2 x = \log\log\max\{e^e, x\} \quad x \geq 0.$$

According to Orey's theorem (see, e.g. Theorem 2.6, Chapter 6 in Revuz (1975)), when a Markov chain $\{X_n\}_{n\geq 0}$ is aperiodic and Harris recurrent, the σ-algebra

$$\mathcal{A} = \bigcap_{n=0}^{\infty} \sigma\{X_k;\ k \geq n\}$$

is a.s. trivial: for every $A \in \mathcal{A}$, $P(A)$ is identically zero or one. Since by definition, ergodicity implies aperiodicity and Harris recurrence, in particular, the strong limiting behavior of the normalized sums in (1.1) does not depend on the choice of the initial distribution of the chain $\{X_n\}_{n\geq 0}$ and with probability one, the sequence in (1.1) is either eventually bounded or unbounded. We say that the law of the iterated logarithm (LIL) holds if the sequence in (1.1) is eventually bounded with probability one.

The previous literature on the LIL for Markov chains can be found in [17], [19], [33] and [43], where the map ξ takes the real values and the regeneration and split chain (in the general state space context) are used as a main tool. The existing problem is how to state our result in terms of the original chain (rather than the split chain), as well as to establish it under the possibly weakest conditions.

We solve this problem in the following steps: In Section III-2 we compare the LIL with the CLT in the real valued context. We give an example and a theorem in this section to show that the CLT does not imply the LIL but basically, the LIL implies the CLT. In Section III-3, we introduce the concept of reproducing kernel Hilbert space to describe the clustering phenomena of the LIL in the Banach valued case. The idea originally comes from its counterpart in the i.i.d. case and some arguments are adapted from [2] and [26]. In Section III-4, we give our main theorem in the vector setting, where all conditions given in the theorem are necessary or nearly necessary for the LIL. The "weak" central limit theorem is one of these conditions and is also used to identify various limits in the LIL. This appears to be natural and reasonable since in the real valued context, the CLT turns out to be necessary for the LIL and is well understood through the results in Chapter II. Another nontrivial condition, which turns out to be necessary for the LIL, is presented in Chapter I as maximal integrability on small sets. Since the resolvent argument developed in Chapter II does not work in the vector setting (Example 6.1), we have to use the split chain construction in the general sense and we approximate the chain by an 1-dependent sequence which satisfies the LIL by an earlier result obtained by the author ([14]). In Section III-5, we give some corollaries and simplified versions of our main theorem under certain more specific conditions. In Section III-6, we make some remarks on the argument used in this chapter. An example is given to show some singular phenomena in the vector setting and a theorem on the law of the large numbers of Marcinkiewicz-Zygmund's type is stated to illustrate how the argument used here can be used to establish some other strong limit theorems for ergodic Markov chains.

III-2. The LIL and CLT

In this section, we assume that ξ takes real values, i.e. $\xi\colon E \longrightarrow \mathbf{R}$ is a measurable function.

Recall that in the real i.i.d. context, the CLT and the LIL are equivalent. It is natural to ask if this is so in the Markovian context. Unfortunately, the answer is "no". Morever, the following example shows that for every $1 < p < 2$, there exists an ergodic Markov chain $\{X_n\}_{n\geq 0}$ and a bounded and mean zero function ξ on the state space E, such that the CLT holds but

$$(2.1) \qquad \limsup_{n\to\infty} \frac{1}{n^{1/p}} S_n = +\infty \quad \text{a.s.}$$

which in particular leads to the violation of the LIL.

Example 2.1. Let $E = \{0, 1, 2, \cdots\}$ and define the subsets I_k, J_k of E as follows:

$$I_0 = [1, 2], \quad I_k = (2^{2k}, 2^{2k+1}] \ (k \geq 1) \quad \text{and} \quad J_k = (2^{2k+1}, 2^{2k+2}] \ (k \geq 0).$$

Fix $1 < p < 2$ and define $\{p_n\}_{n\geq 1}$ as follows:

$$p_n = \begin{cases} \left(\frac{n}{n+1}\right)^{2p} & n \in \bigcup_{k=0}^{\infty} I_k \\ 1 - \frac{1}{n^2} & n \in \bigcup_{k=0}^{\infty} J_k. \end{cases}$$

One can easily check that there exists constants $\lambda_1, \lambda_2 > 0$ such that

(2.2) $$\lambda_1 \left(\frac{1}{n}\right)^p \leq p_1 p_2 \cdots p_n \leq \lambda_2 \left(\frac{1}{n}\right)^p$$

hold for all n. Consequently,

$$\sum_{n=1}^{\infty} (p_1 \cdots p_n) < +\infty.$$

Define the Markov transition $\{p_{ij}\}_{i,j \in E}$ by

$$p_{ij} = \begin{cases} 1 - p_{i+1} & j = 0 \\ p_{i+1} & j = i+1 \\ 0 & \text{else.} \end{cases}$$

Then, $\{p_{ij}\}_{i,j \in E}$ is ergodic with the invariant distribution π given by

$$\begin{cases} \pi(0) = \left(1 + \sum_{n=1}^{\infty} (p_1 \cdots p_n)\right)^{-1} \\ \pi(k) = \left(1 + \sum_{n=1}^{\infty} (p_1 \cdots p_n)\right)^{-1} p_1 \cdots p_k \quad (k \geq 1) \end{cases}$$

Let $\{X_n\}_{n \geq 0}$ be a Markov chain with the transition $\{p_{ij}\}_{i,j \in E}$ and define $\xi: E \longrightarrow \mathbf{R}$ in the following way:

$$\xi(j) = 0 \quad \text{when } j \in \bigcup_{k=0}^{\infty} I_k;$$

for $j \in J_k$ $(k \geq 0)$, define

$$\xi(j) = \begin{cases} 1 & j \in \left(2^{2k+1}, 2^{2k+1} + 2^{2k}\right] \\ -1 & j \in \left(2^{2k+1} + 2^{2k}, 2^{2k+2}\right]; \end{cases}$$

and, define $\xi(0)$ in such a way that

(2.3) $$\sum_{i \in E} \pi(i) \xi(i) = 0.$$

Then, ξ is bounded and centered. Define

$$\tau_0 = \inf\{n \geq 1; X_n = 0\}.$$

To prove that the CLT holds, by Theorem II-2.1 we need only to prove that

(2.4) $\quad E_0\Big(\sum_{k=1}^{\tau_0} \xi(X_k)\Big) = 0 \quad \text{and} \quad E_0\Big(\sum_{k=1}^{\tau_0} \xi(X_k)\Big)^2 < +\infty.$

The first assertion follows from (2.3) and Theorem 10.4.9 in Meyn-Tweedie (1993). We now prove the second one. Let

$$s_n = \sum_{k=0}^{n-1} \xi(k) \quad n = 1, 2, \cdots.$$

Then,

(i) $s_n = \xi(0)$ when $n \in \bigcup_{k=0}^{\infty} I_k$;

(ii) there exists $M > 0$ such that $|s_n| \leq Mn$ for all n;

(iii) $s_{2^{2k+1}+2^{2k}+1} = \xi(0) + 2^{2k} \quad k \geq 0$.

Note that that

$$\sum_{k=1}^{\tau_0} \xi(X_k) = \xi(0) + \sum_{k=1}^{\tau_0-1} \xi(X_k) = \xi(0) + \sum_{k=1}^{\tau_0-1} \xi(k) = s_{\tau_0} \quad \text{a.s.} \quad P_0.$$

Hence,

$$E_0\Big(\sum_{k=1}^{\tau_0} \xi(X_k)\Big)^2 = E_0 s_{\tau_0}^2 = \sum_{n=1}^{\infty} s_n^2 P_0\{\tau_0 = n\}$$

$$\leq |\xi(0)|^2 + M \sum_k \sum_{n \in J_k} n^2 P_0\{\tau_0 = n\}$$

$$= |\xi(0)|^2 + M \sum_k \sum_{n \in J_k} n^2 p_1 \cdots p_{n-1}(1-p_n)$$

$$\leq |\xi(0)|^2 + M\lambda_2 \sum_k \sum_{n \in J_k} n^2 \frac{1}{(n-1)^p} \cdot \frac{1}{n^2} < +\infty.$$

To prove (2.1), first we claim that $E_0\tau_0^p = +\infty$. In fact,

$$E_0\tau_0^p \geq \sum_k \sum_{n \in I_k} n^p P\{\tau_0 = n\}$$

$$= \sum_k \sum_{n \in I_k} n^p \cdot p_1 \cdots p_{n-1}(1-p_n)$$

$$\geq \lambda_1 \sum_k \sum_{n \in I_k} n^p \frac{1}{n^p}\left(1 - \left(\frac{n}{n+1}\right)^{2p}\right).$$

Note that

$$1 - \left(\frac{n}{n+1}\right)^{2p} \sim \frac{2p}{n} \quad (n \to \infty)$$

and that

$$\sum_{n \in I_k} \frac{1}{n} \geq \log \frac{2^{2k+1}+1}{2^{2k}+1} \longrightarrow \log 2 \quad (k \to \infty).$$

Hence our claim holds.

Define

$$\begin{cases} \tau_0(1) = \tau_0 \\ \tau_0(k+1) = \inf\{n > \tau_0(k); \ X_n = 0\} \quad (k \geq 1) \end{cases}$$

Then, $\{\tau_0(k+1) - \tau_0(k)\}_{k \geq 1}$ is an i.i.d. random sequence with the common distribution:

$$P_0\{\tau_0 = j\} \quad j = 1, 2, \cdots.$$

By the law of the large numbers,

(2.5) $$\lim_{n \to \infty} \frac{\tau_0(n)}{n} = E_0\tau_0 \quad \text{a.s.}$$

Note that

$$\sum_{n=1}^{\infty} P\{\tau_0(n+1) - \tau_0(n) \geq Ln^{1/p}\} = \sum_{n=1}^{\infty} P_0\{\tau_0 \geq Ln^{1/p}\} = L^{-p}E_0\tau_0^p = +\infty$$

for all $L > 0$. Applying the Borel-Cantelli lemma gives

$$\limsup_{n \to \infty} \frac{\tau_0(n+1) - \tau_0(n)}{n^{1/p}} = +\infty \quad \text{a.s.}$$

Given $N > 0$, therefore, almost surely there exists a subsequence $\{n_k(\omega)\}_{k \geq 1}$ of positive integers such that

$$\tau_0(n_k + 1) - \tau_0(n_k) > (Nn_k)^{1/p} \quad k \geq 1.$$

Let $m_k = \left[(2p \log 2)^{-1} \log(Nn_k)\right] - 1$ $(k \geq 1)$. Then,

$$l_k \equiv \tau_0(n_k) + 2^{2m_k+1} + 2^{2m_k} < \tau_0(n_k + 1) \quad \text{a.s.}$$

Therefore,

$$S_{l_k+1} - S_{\tau_0(n_k)+1} = \sum_{j=\tau_0(n_k)+1}^{l_k} \xi(X_j)$$

$$= \sum_{j=\tau_0(n_k)+1}^{l_k} \xi(j - \tau_0(n_k)) = \sum_{j=1}^{2^{2m_k+1}+2^{2m_k}} \xi(j)$$

$$= s_{2^{2m_k+1}+2^{2m_k}+1} - \xi(0) = 2^{2m_k} \geq \frac{1}{8}(Nn_k)^{1/p} \quad \text{a.s.}$$

Consequently,

$$\liminf_{k \to \infty} n_k^{-1/p} \left(S_{l_k+1} - S_{\tau_0(n_k)+1}\right) \geq \frac{1}{8} N^{1/p} \quad \text{a.s.}$$

Note that $\left\{\sum_{j=\tau_0(k)+1}^{\tau_0(k+1)} \xi(X_j)\right\}_{k \geq 1}$ is an i.i.d. sequence with the common distribution:

$$\mathcal{L}_{P_0}\left(\sum_{j=1}^{\tau_0} \xi(X_j)\right).$$

In view of (2.4), applying the Marcikinewicz-Zygmund's law of the large numbers to this sequence yields:

$$\lim_{k \to \infty} \frac{1}{k^{1/p}} S_{\tau_0(k)+1} = 0 \quad \text{a.s.}$$

Hence, we have that

$$\liminf_{k\to\infty} n_k^{-1/p} S_{l_k+1} \geq \frac{1}{8} N^{1/p} \quad \text{a.s.}$$

From (2.5), $l_k \sim n_k E_0 \tau_0$ a.s. as $k \to \infty$. Therefore,

$$\limsup_{n\to\infty} \frac{1}{n^{1/p}} S_n \geq \frac{1}{8} N^{1/p} (E_0 \tau_0)^{-1/p} \quad \text{a.s.}$$

Letting $N \longrightarrow +\infty$ gives (2.1). ∎

However, the following theorem shows that the LIL implies the CLT in a very general setting.

Theorem 2.2. *Let $\{X_n\}_{n\geq 0}$ be an ergodic Markov chain and let ξ satisfy*

(2.6) $$\int \xi^2(x) \pi(dx) < +\infty.$$

Assume that

(2.7) $$\limsup_{n\to\infty} \frac{|S_n|}{\sqrt{2nL_2 n}} < +\infty \quad a.s.$$

Then, there exists $\sigma \geq 0$ such that

(2.8) $$S_n/\sqrt{n} \longrightarrow N(0, \sigma^2) \quad \text{in distribution}.$$

Morever, we have

(2.9) $$\limsup_{n\to\infty} \frac{|S_n|}{\sqrt{2nL_2 n}} = \sigma \quad a.s.,$$

(2.10) $$\limsup_{n\to\infty} \frac{S_n}{\sqrt{2nL_2 n}} = \sigma \quad a.s.$$

and

(2.11) $$\liminf_{n\to\infty} \frac{S_n}{\sqrt{2nL_2n}} = -\sigma \quad \text{a.s.}$$

Remark 2.3. From the following proof, we will see that when $\{X_n\}_{n\geq 0}$ has a small set of order 1, the conclusion of Theorem 2.2 stands even without (2.6).

Proof. By Orey's theorem (Theorem 2.6, p167, [40]), there exists a constant $\sigma \geq 0$ such that (2.9) holds. We need only to prove that such $\sigma \geq 0$ satisfies (2.8) and (2.10) since (2.11) follows from (2.10) with $-\xi$ instead of ξ. We proceed step by step.

Step 1. We prove (2.8) and (2.10) under the additional assumption that $\{X_n\}_{n\geq 0}$ has an atom α. Let the regeneration times $\{\tau_\alpha(k)\}_{k\geq 1}$ be given in (I-3.7). By the law of the large numbers for i.i.d. sequences,

(2.12) $$\lim_{n\to\infty} \frac{\tau_\alpha(n)}{n} = E_\alpha \tau_\alpha = \frac{1}{\pi(\alpha)} \quad \text{a.s.}$$

where the second equality follows from Theorem 10.4.9 in Meyn-Tweedie (1993).

Notice that (2.9) implies that

$$\limsup_{n\to\infty} \frac{|S_{\tau_\alpha(n)+1}|}{\sqrt{2(\tau_\alpha(n)+1)L_2(\tau_\alpha(n)+1)}} \leq \sigma \quad \text{a.s.}$$

From (2.12),

(2.13) $$\limsup_{n\to\infty} \frac{|S_{\tau_\alpha(n)+1}|}{\sqrt{2nL_2n}} \leq \left(\pi(\alpha)\right)^{-1/2} \sigma \quad \text{a.s.}$$

Again, (2.9) implies that

$$\limsup_{n\to\infty} \max_{k\leq n} |S_k|/\sqrt{2nL_2n} = \sigma \quad \text{a.s.}$$

Similarly,

$$\limsup_{n\to\infty} \max_{k\leq \tau_\alpha(n)+1} |S_k|/\sqrt{2nL_2 n} \leq (\pi(\alpha))^{-1/2}\sigma \quad \text{a.s.}$$

By the triangle inequality we have

$$\limsup_{n\to\infty} \max_{\tau_\alpha(n)<k\leq \tau_\alpha(n+1)} \left|\sum_{j=\tau_\alpha(n)+1}^{k} \xi(X_j)\right|/\sqrt{2nL_2 n} \leq 2(\pi(\alpha))^{-1/2}\sigma \quad \text{a.s.}$$

As is well known, the blocks $\{B_n\}_{n\geq 1}$ given by

$$B_n = \{X_{\tau_\alpha(n)+1}, \cdots, X_{\tau_\alpha(n+1)}\} \quad n=1,2,\cdots$$

are i.i.d. with the common distribution:

$$\mathcal{L}_{P_\alpha}(\{X_1,\cdots,X_{\tau_\alpha}\}).$$

By the Borel-Cantelli lemma, there exists a constant $M > 0$ such that

$$\sum_n P_\alpha\left\{\max_{1\leq k\leq \tau_\alpha} \left|\sum_{j=1}^{k} \xi(X_j)\right| \geq M\sqrt{2nL_2 n}\right\} < +\infty,$$

which is equivalent to

$$\sum_n P_\alpha\left\{\max_{1\leq k\leq \tau_\alpha} \left|\sum_{j=1}^{k} \xi(X_j)\right| \geq \epsilon\sqrt{2nL_2 n}\right\} < +\infty \quad \forall \epsilon > 0.$$

Applying the Borel-Cantelli lemma again gives

$$(2.14) \quad \limsup_{n\to\infty} \max_{\tau_\alpha(n)<k\leq \tau_\alpha(n+1)} \left|\sum_{j=\tau_\alpha(n)+1}^{k} \xi(X_j)\right|/\sqrt{2nL_2 n} = 0 \quad \text{a.s.}$$

For each $n \geq 1$, let $i_\alpha(n)$ be given in (I-3.6). Then, on the event $\{i_\alpha(n) \geq 1\}$,

$$(2.15) \quad \tau_\alpha(i_\alpha(n)) \leq n-1 < \tau_\alpha(i_\alpha(n)+1) \quad \text{a.s.}$$

Consider (II-2.6). By (2.15) we have

$$\left|\sum_{j=l_\alpha(n)+1}^{n-1} \xi(X_j)\right| \le \max_{\tau_\alpha(i_\alpha(n))<k\le\tau_\alpha(i_\alpha(n)+1)} \left|\sum_{j=\tau_\alpha(i_\alpha(n))+1}^{k} \xi(X_j)\right| \quad \text{a.s.}$$

From (2.14),

$$\lim_{n\to\infty}\left|\sum_{j=l_\alpha(n)+1}^{n-1} \xi(X_j)\right| \Big/ \sqrt{2i_\alpha(n)L_2 i_\alpha(n)} = 0 \quad \text{a.s.}$$

By (2.12) we have that

(2.16) $$\lim_{n\to\infty}\left|\sum_{j=l_\alpha(n)+1}^{n-1} \xi(X_j)\right| \Big/ \sqrt{2nL_2n} = 0 \quad \text{a.s.}$$

Notice that

$$|S_{n\wedge(\tau_\alpha+1)}| \le \sum_{k=0}^{\tau_\alpha} |\xi(X_k)|.$$

Therefore, it follows from (II-2.6) and (2.9) that

$$\limsup_{n\to\infty}\left|\sum_{k=1}^{i_\alpha(n)-1} \xi_k\right| \Big/ \sqrt{2nL_2n} = \sigma \quad \text{a.s.}$$

Using (2.12) again gives

$$\limsup_{n\to\infty}\left|\sum_{k=1}^{i_\alpha(n)-1} \xi_k\right| \Big/ \sqrt{2i_\alpha(n)L_2 i_\alpha(n)} = \left(\pi(\alpha)\right)^{-1/2}\sigma \quad \text{a.s.}$$

Consequently,

$$\limsup_{n\to\infty}\left|\sum_{k=1}^{n-1} \xi_k\right| \Big/ \sqrt{2nL_2n} \ge \left(\pi(\alpha)\right)^{-1/2}\sigma \quad \text{a.s.}$$

Notice that

(2.17) $$S_{\tau_\alpha(n)+1} = S_{\tau_\alpha+1} + \sum_{k=1}^{n-1} \xi_k \quad n=1,2,\cdots.$$

Hence, by (2.13) we have

$$\limsup_{n\to\infty}\Big|\sum_{k=1}^{n-1}\xi_k\Big|\Big/\sqrt{2nL_2n} \le \big(\pi(\alpha)\big)^{-1/2}\sigma \quad \text{a.s.}$$

Therefore

$$\limsup_{n\to\infty}\Big|\sum_{k=1}^{n-1}\xi_k\Big|\Big/\sqrt{2nL_2n} = \big(\pi(\alpha)\big)^{-1/2}\sigma \quad \text{a.s.}$$

By the converse of the Hartman-Wintner LIL, the statement (iii) in Theorem II-2.1 holds and σ satisfies (II-2.1) in Theorem II-2.1. Therefore, (2.8) follows from Theorem II-2.1. We now prove (2.10). Clearly, by (2.9) we need only to prove that

$$(2.18) \qquad \limsup_{n\to\infty} \frac{S_n}{\sqrt{2nL_2n}} \ge \sigma \quad \text{a.s.}$$

Applying the Hartman-Wintner LIL to $\{\xi_k\}_{k\ge 1}$ again yields that

$$\limsup_{n\to\infty} \sum_{k=1}^{n-1}\xi_k\Big/\sqrt{2nL_2n} = \big(\pi(\alpha)\big)^{-1/2}\sigma \quad \text{a.s.}$$

By (2.17), this is equivalent to

$$\limsup_{n\to\infty} \frac{S_{\tau_\alpha(n)+1}}{\sqrt{2nL_2n}} = \big(\pi(\alpha)\big)^{-1/2}\sigma \quad \text{a.s.}$$

By (2.12), this is equivalent to

$$\limsup_{n\to\infty} \frac{S_{\tau_\alpha(n)+1}}{\sqrt{2(\tau_\alpha(n)+1)L_2(\tau_\alpha(n)+1)}} = \sigma \quad \text{a.s.}$$

which implies (2.18).

Step 2. We now assume additionally that $\{X_n\}_{n\ge 0}$ has a small set C of order 1. In this case $\{X_n\}_{n\ge 0}$ can be embedded by the splitting method into an enlarged probability space together with a sequence $\{Y_n\}_{n\ge 0}$ of $\{0,1\}$

valued random variables, such that $\{(X_n, Y_n)\}_{n\geq 0}$ is an ergodic Markov chain with an atom $\alpha^* = C \times \{1\}$. Therefore the result proved at Step 1 can be passed to this step simply by viewing ξ as a function defined on $E \times I$.

Step 3. We consider the resolvent chain in the general context. Namely, for each $0 < t < 1$ let the random sequences $\{\delta_k\}_{k\geq 1}$ and $\{T_k\}_{k\geq 0}$ be given in (II-3.3) and (II-3.4), respectively and, let $\{X_n^t\}_{n\geq 0}$ be a resolvent chain with the transition P_t given in (I-3.6). Without loss of generality, let π be the initial distribution of $\{X_n\}_{n\geq 0}$. First we claim that

$$(2.19) \quad \limsup_{n\to\infty} \left|\sum_{k=1}^{n} (\delta_k - (1-t))\xi(X_k)\right| / \sqrt{2nL_2n}$$
$$= \sqrt{t(1-t)} \left(\int \xi^2(x)\pi(dx)\right)^{1/2} \quad \text{a.s.}$$

Recall (see, e.g. Appendix IV, [27]) that a strictively stationary random sequence $\{Z_n\}_{n\geq 1}$ is called ergodic if the σ-algebra

$$\mathcal{F} = \bigcap_{n=1}^{\infty} \sigma\{Z_k;\ k \geq n\}$$

is a.s. trivial: for every $A \in \mathcal{F}$, $P(A)$ is either zero or one. While this definition is slightly different from the one we give for Markov chains, one can find some connections through Orey's theorem. For our needs here, we prove that the strictively stationary sequence $\{(\delta_n - (1-t))\xi(X_n)\}_{n\geq 1}$ of martingale differences is ergodic with the definition given for stationary sequences. Indeed, by Orey's theorem (Theorem 2.6, p167, [40]) the σ-algebra

$$\mathcal{A} = \bigcap_{n=0}^{\infty} \sigma\{X_k;\ k \geq n\}$$

is a.s. trivial and by the Kolmogorov 0-1 law, so is the σ-algebra $\mathcal{D}$ given by

$$\mathcal{D} = \bigcap_{n=1}^{\infty} \sigma\{\delta_k;\ k \geq n\}.$$

Therefore, the ergodicity we claim for the sequence $\{(\delta_n - (1-t))\xi(X_n)\}_{n \geq 1}$ follows from the fact that the σ-algebra $\mathcal{A} \vee \mathcal{D}$ is a.s. trivial and the relation

$$\bigcap_{n=1}^{\infty} \sigma\{(\delta_k - (1-t))\xi(X_k);\ k \geq n\} \subset \mathcal{A} \vee \mathcal{D}.$$

One can also verify that

$$E\Big((\delta_1 - (1-t))\xi(X_1)\Big)^2 = t(1-t)\int \xi^2(x)\pi(dx).$$

Therefore, (2.19) follows from the Stout's LIL ([41]) for stationary ergodic martingale differences.

From (2.9) and (2.19),

(2.20)
$$(1-t)\sigma - \sqrt{t(1-t)}\Big(\int \xi^2(x)\pi(dx)\Big)^{1/2}$$
$$\leq \limsup_{n\to\infty} \Big|\sum_{k=1}^{n} \delta_k \xi(X_k)\Big|/\sqrt{2nL_2 n}$$
$$\leq (1-t)\sigma + \sqrt{t(1-t)}\Big(\int \xi^2(x)\pi(dx)\Big)^{1/2} \qquad \text{a.s.}$$

In particular, we have

$$\limsup_{n\to\infty} \Big|\sum_{k=1}^{T_n} \delta_k \xi(X_k)\Big|/\sqrt{2T_n L_2 T_n}$$
$$\leq (1-t)\sigma + \sqrt{t(1-t)}\Big(\int \xi^2(x)\pi(dx)\Big)^{1/2} \qquad \text{a.s.}$$

By the law of the large numbers for i.i.d. sequences,

(2.21)
$$\lim_{n\to\infty} \frac{T_n}{n} = (1-t)^{-1} \qquad \text{a.s.}$$

Therefore,

(2.22)
$$\limsup_{n\to\infty} \Big|\sum_{k=1}^{T_n} \delta_k \xi(X_k)\Big|/\sqrt{2nL_2 n}$$
$$\leq \sqrt{1-t}\,\sigma + \Big(t\int \xi^2(x)\pi(dx)\Big)^{1/2} \qquad \text{a.s.}$$

On the other hand, by the definitions of $\{\delta_k\}_{k\geq 1}$ and $\{T_k\}_{k\geq 0}$,

$$\max_{k\leq n}\Big|\sum_{j=1}^{T_k}\delta_j\xi(X_j)\Big| = \max_{k\leq T_n}\Big|\sum_{j=1}^{k}\delta_j\xi(X_j)\Big| \geq \Big|\sum_{k=1}^{n}\delta_k\xi(X_k)\Big| \quad \text{a.s.}$$

for each $n \geq 1$. Using (2.20) again gives

(2.23)
$$\begin{aligned}
&\limsup_{n\to\infty}\Big|\sum_{k=1}^{T_n}\delta_k\xi(X_k)\Big|\Big/\sqrt{2nL_2n} \\
&= \limsup_{n\to\infty}\max_{k\leq n}\Big|\sum_{j=1}^{T_k}\delta_j\xi(X_j)\Big|\Big/\sqrt{2nL_2n} \\
&\geq (1-t)\sigma - \sqrt{t(1-t)}\Big(\int\xi^2(x)\pi(dx)\Big)^{1/2} \quad \text{a.s.}
\end{aligned}$$

By Orey's theorem (Theorem 2.6, p167, [40]) and Lemma II-3.3, (2.22) and (2.23) imply that for each $0 < t < 1$, there exists a constant $\sigma_t \geq 0$ such that

(2.24) $$\limsup_{n\to\infty}\Big|\sum_{k=0}^{n-1}\xi(X_k^t)\Big|\Big/\sqrt{2nL_2n} = \sigma_t \quad \text{a.s.}$$

and

(2.25) $$\begin{aligned}
(1-t)\sigma - \sqrt{t(1-t)}\Big(\int\xi^2(x)\pi(dx)\Big)^{1/2} \\
\leq \sigma_t \leq \sqrt{1-t}\sigma + \Big(t\int\xi^2(x)\pi(dx)\Big)^{1/2}.
\end{aligned}$$

Applying the conclusion achieved at Step 2 to the chain $\{X_n^t\}_{n\geq 0}$ gives

(2.26) $$\sum_{k=0}^{n-1}\xi(X_k^t)/\sqrt{n} \longrightarrow N(0,\sigma_t^2) \quad \text{in distribution}$$

and

(2.27) $$\limsup_{n\to\infty}\sum_{k=0}^{n-1}\xi(X_k^t)/\sqrt{2nL_2n} = \sigma_t \quad \text{a.s.}$$

From (2.25)

(2.28) $$\lim_{t\to 0^+} \sigma_t = \sigma.$$

Hence, applying Lemma II-3.4 gives (2.8). To prove (2.10), it is enough to show that (2.18) holds in the general context. In view of (2.21), it follows from Lemma II-3.3 that (2.27) is equivalent to

$$\limsup_{n\to\infty} \sum_{k=1}^{T_n} \delta_k \xi(X_k)/\sqrt{2T_n L_2 T_n} = \sqrt{1-t}\cdot \sigma_t \quad \text{a.s.}$$

from which we have

$$\limsup_{n\to\infty} \sum_{k=1}^{n} \delta_k \xi(X_k)/\sqrt{2nL_2 n} \geq \sqrt{1-t}\cdot \sigma_t \quad \text{a.s.}$$

This, together with (2.20) and the triangle inequality, implies

$$\limsup_{n\to\infty} \frac{S_n}{\sqrt{2nL_2 n}} \geq (1-t)^{-1/2}\sigma_t - \left(\frac{t}{1-t}\int \xi^2(x)\pi(dx)\right)^{1/2} \quad \text{a.s.}$$

By (2.28), letting $t \longrightarrow 0^+$ gives (2.18). ∎

We end this section with the following comment: Theorem 2.2 tells us that once the LIL is satisfied, the limit value in the LIL is determined by the limit variance parameter σ^2 in the CLT. Therefore, the limit value in the LIL can be identified by using Theorem II-3.1. Namely, if the LIL is satisfied and if the conditions (i) and (ii') in Theorem II-3.1 are fulfilled, then (2.9), (2.10) and (2.11) hold with the parameter σ given by

$$\sigma = \left\{ \int \xi^2(x)\pi(dx) + 2\int \sum_{n=1}^{\infty} \xi(x)P^n\xi(x)\pi(dx) \right\}^{1/2}.$$

However, the problem still exists: How can the LIL be satisfied? We leave this to Section III-4 where we solve it in the vector setting.

III-3. Some remarks on the limit set

Let B be a separable Banach space with norm $||\cdot||$ and topological dual B^* whose closed unit ball is denoted by B_1^*. Unless stated otherwise, in the rest of this Chapter, ξ is a measurable map from E to B. For a fixed ergodic Markov chain $\{X_n\}_{n\geq 0}$, we say ξ satisfies the **weak central limit theorem** (write $\xi \in WCLT$ for short) if for each $f \in B^*$, there is $\sigma_f^2 \geq 0$ such that

$$(3.1) \qquad f(S_n/\sqrt{n}) \longrightarrow N(0, \sigma_f^2) \text{ in distribution}$$

and we call $\{\sigma_f^2;\ f \in B^*\}$ the **Gaussian covariance function** (g.c.f.) associated with ξ. From definition, $\{\sigma_f^2;\ f \in B^*\}$ is a quadratic form on B^*: for any $a \in \mathbf{R}$ and $f, g \in B^*$,

$$(3.2) \qquad \begin{cases} \sigma_{af} = |a| \cdot \sigma_f, \quad \sigma_{f+g} \leq \sigma_f + \sigma_g, \\ \sigma_{f+g}^2 + \sigma_{f-g}^2 = 2(\sigma_f^2 + \sigma_g^2). \end{cases}$$

Indeed, for any $a, b \in \mathbf{R}$ and $f, g \in B^*$,

$$af(S_n/\sqrt{n}) + bg(S_n/\sqrt{n}) \longrightarrow N(0, \sigma_{af+bg}^2) \text{ in distribution.}$$

Therefore, there exists a centered normal distribution $\gamma_{f,g}$ on $\mathbf{R}^2$ such that

$$\left(f(S_n/\sqrt{n}),\ g(S_n/\sqrt{n})\right) \longrightarrow \gamma_{f,g} \text{ in distribution.}$$

Consequently,

$$\sigma_{af+bg}^2 = \int (as+bt)^2 \gamma_{f,g}(d(s,t)), \quad \forall a, b \in \mathbf{R}^2,$$

from which one can easily see how (3.2) follows.

Recall that in the independent ([26]) and m-dependent ([14]) contexts, the concept of reproducing kernel Hilbert space is introduced to describe the cluster behavior of the LIL in the vector setting. For the same reason, we introduce it to the Markovian context. A subspace H of B is called the **reproducing kernel Hilbert space** generated by the g.c.f. $\{\sigma_f^2;\ f \in B^*\}$, if there exist a norm $||\cdot||_H$ on H such that

(1). under $||\cdot||_H$, H is a Hilbert space;

(2). there exists a linear continuous operator $S: B^* \longrightarrow (H, ||\cdot||_H)$ such that the image $Im(S)$ of S is dense in $(H, ||\cdot||_H)$ and

$$\text{(3.3)} \qquad ||Sf||_H^2 = \sigma_f^2 \quad \forall f \in B^*,$$

$$\text{(3.4)} \qquad (Sf, Sg)_H = f(Sg) = g(Sf) \quad \forall f, g \in B^*,$$

where $(\cdot,\cdot)_H$ is the inner product w.r.t. $||\cdot||_H$. The operator S is called the **covariance operator** associated with $\{\sigma_f^2;\ f \in B^*\}$.

In view of the construction of the reproducing kernel Hilbert space in the i.i.d. context (see Lemma 2.1 in [26]), the following questions arise naturally about our definition:

Question 1. Does the g.c.f. $\{\sigma_f^2;\ f \in B^*\}$ generate a reproducing kernel Hilbert space?

Question 2. Are the reproducing kernel Hilbert space $(H, ||\cdot||_H)$ and the covariance operator S, if they exist, uniquely determined by the g.c.f. $\{\sigma_f^2;\ f \in B^*\}$?

Question 3. Like in the classical case, some properties of the reproducing kernel Hilbert space are needed in order to connect it to the cluster behavior of the LIL. Can these properties be derived from the definition?

We now answer Question 2 and 3 by giving the following:

Lemma 3.1. *Let $(H, ||\cdot||_H)$ and S be, respectively, the reproducing kernel Hilbert space and covariance operator associated with the g.c.f. $\{\sigma_f^2;\ f \in B^*\}$. Then:*

(i) We have

$$(3.5) \qquad \sigma \equiv \sup_{f \in B_1^*} \sigma_f = ||S|| < +\infty,$$

$$(3.6) \qquad ||x|| \leq \sigma ||x||_H \quad \forall x \in H.$$

(ii) $(H, ||\cdot||_H)$ is uniquely determined by $\{\sigma_f^2;\ f \in B^*\}$ in the following way:

$$(3.7)\ H = \{x \in B;\ \sup_{\sigma_f \leq 1} f(x) < +\infty\}\ \text{and}\ ||x||_H = \sup_{\sigma_f \leq 1} f(x) \quad \forall x \in H,$$

and the covariance operator S is uniquely defined in the sense that for any continuous linear operator $S': B^* \longrightarrow (H, ||\cdot||_H)$ with dense image $Im(S')$ in H and

$$(3.8) \qquad (S'f, S'g)_H = f(S'g) = g(S'f) \quad \forall f, g \in B^*$$

we have $S' = S$.

(iii) If K is the closed unit ball of $(H, ||\cdot||_H)$, then K is a closed symmetric convex subset of B and for each $f \in B^*$,

$$(3.9) \qquad \sup_{x \in K} f(x) = \sigma_f.$$

Further, K is a compact subset of B iff the g.c.f. σ_f^2 is weak-star sequentially continuous, i.e., iff for all sequences $\{f_k\}_{k \geq 1}$ in B^* such that $f_k \to f$ in the weak-star sense, we have $\sigma_{f_k} \to \sigma_f$.

Proof. Clearly, (3.5) follows from (3.3) and continuity of S. Since $Im(S)$ is dense in H, (i) holds if (3.6) is true whenever $x = Sg$ for some $g \in H$. By (3.3) and (3.4),

$$\|Sg\| = \sup_{f \in B_1^*} f(Sg) = \sup_{f \in B_1^*} (Sf, Sg)_H$$
$$\leq \sup_{f \in B_1^*} \|Sf\|_H \cdot \|Sg\|_H = \sup_{f \in B_1^*} \sigma_f \cdot \|Sg\|_H = \sigma \|Sg\|_H.$$

Given $g \in B^*$,

$$\sup_{\sigma_f \leq 1} f(Sg) = \sup_{\sigma_f \leq 1} (Sf, Sg)_H = \sup_{\|f\|_H \leq 1} (Sf, Sg)_H = \|Sg\|_H < +\infty.$$

Hence, the density of $Im(S)$ implies the second equality in (3.7) and

(3.10) $$H \subset \{x \in B; \sup_{\sigma_f \leq 1} f(x) < +\infty\}.$$

On the other hand, suppose $x_o \in B$ satisfies

(3.11) $$\sup_{\sigma_f \leq 1} f(x_o) < +\infty.$$

Consider the linear function h_{x_o} on $Im(S)$ given by

$$h_{x_o}(Sf) = f(x_o) \quad f \in B^*.$$

First we show that h_{x_o} is well defined: For any $f \in B^*$ with $Sf = 0$, by (3.2) we have $\sigma_{af} = 0$ for all $a > 0$. Consequently,

$$a|f(x_o)| \leq \sup_{\sigma_g \leq 1} g(x_o) < +\infty \quad \forall a > 0,$$

which implies $f(x_o) = 0$. From (3.11) and the density of $Im(S)$ in H, h_{x_o} uniquely determine a continuous linear function on H. By the Fréchet-Riesz representation theorem, there exists unique $x'_o \in H$ such that

$$(Sf, x'_o)_H = h_{x_o}(Sf) = f(x_o) \quad f \in B^*.$$

On the other hand, from the density of $Im(S)$ in H and (3.4) we have

(3.12) $$(Sf, x)_H = f(x) \quad \forall x \in H.$$

Hence, $f(x'_o) = (Sf, x'_o)_H = f(x_o)$ for all $f \in B^*$ and thus $x_o = x'_o \in H$. Therefore, we have proved

(3.13) $$H \supset \{x \in B;\ \sup_{\sigma_f \leq 1} f(x) < +\infty\}.$$

Combining (3.10) and (3.13) gives the first equality in (3.7).

We now prove the assertion regarding the uniqueness of S. From (3.8) and (3.12),

$$(Sf, S'g)_H = f(S'g) = (S'f, S'g)_H \quad f, g \in B^*.$$

By the density of $Im(S')$ in H, this implies $Sf = S'f$ for all $f \in B^*$.

We now come to the proof of (iii). The symmetry and convexity of K are obvious. To show that K is closed in B, it is enough to prove that $||\cdot||_H$ (where we define $||x||_H = +\infty$ if $x \notin H$) is lower semi-continuous, i.e., for all sequences $\{x_k\}_{k \geq 1}$ in B with $x_k \to x$ w.r.t. the Banach norm $||\cdot||$, we have

(3.14) $$\liminf_{k \to \infty} ||x_k||_H \geq ||x||_H.$$

In fact, from (3.7) we have $||x_k||_H \geq f(x_k)$ for each $f \in B^*$ with $\sigma_f \leq 1$. Consequently,

$$\liminf_{k \to \infty} ||x_k||_H \geq f(x)$$

for each $f \in B^*$ with $\sigma_f \leq 1$. Using (3.7) again gives (3.14).

It is easy to see how (3.3) and (3.12) imply (3.9). It remains to prove the assertions regarding compactness of K. The argument we use in the "if" part is similar to that in [26]: Let $F = \{f \in B^*;\ ||Sf||_H = 0\}$ and consider the quotient space B^*/F. Under the norm $\|Sf\|_H$, B^*/F becomes an inner product space. Let $\overline{B^*/F}$ denote the completion of B^*/F and let $A\colon B^* \longrightarrow \overline{B^*/F}$ be the canonical projection on B^*. From (3.3) and (3.5), one can see that A is a continuous linear operator. Let $A^*\colon \overline{B^*/F} \longrightarrow B^{**}$ be the Banach dual of A. One can see that A^* maps $\overline{B^*/F}$ into B rather than only B^{**}. Morever, K is the image of the closed unit ball of $\overline{B^*/F}$ under the map A^*. According to Alaoglu's theorem, B_1^* is weak-star compact. Hence, the weak-star sequential continuity of σ_f^2 implies that A is a compact operator and thus so is A^* (see Dunford-Schwartz (1964), p.485). Therefore, K is compact in B.

Conversely, suppose K is compact in B and let $\{f_k\}_{k\geq 1}$ be a sequence in B^* such that $f_k \to 0$ in the weak-star sense. By the principle of uniform boundedness,

$$\sup_{k\geq 1} \|f_k\| < +\infty.$$

Hence the functions $\{f_k(x)\}_{k\geq 1}$ are equicontinuous on K (w.r.t. the Banach norm $\|\cdot\|$). By the Arzela-Ascoli theorem, the sequence $\{f_k(x);\ x \in K\}_{k\geq 1}$ is relatively compact in $C(K)$ with the supremum norm. Note that $f_k \to 0$ pointwise. Using (3.9) gives

$$\sigma_{f_k} = \sup_{x\in K} f_k(x) \longrightarrow 0 \quad (k \to \infty).$$

Therefore, σ_f^2 is weak-star continuous. ∎

Concerning Problem 1, we don't know the answer if only the WCLT is assumed. But we can prove the existence of the reproducing kernel Hilbert space under fairly weak conditions, which are enough for our needs in this work. In the following discussion, we assume $\xi \in WCLT$ with the g.c.f. $\{\sigma_f^2;\ f \in B^*\}$.

1. Suppose $\{X_n\}_{n\geq 0}$ has a small set C of order 1, i.e., the minorization

$$(3.15) \qquad P \geq bI_C \otimes \nu$$

holds for some $b > 0$ and some probability measure ν on $(E, \mathcal{E})$. Consider the split chain $\tilde{P}$ (in the classical sense) generated by the minorization (3.15). Let the regeneration times $\{\tau(k)\}_{k\geq 0}$ be given in (I-2.19) (with $m = 1$). Applying Theorem II-2.2 to the function $f \circ \xi$ gives

$$(3.16) \quad \tilde{E}_{\nu^*} f\bigg(\sum_{k=0}^{\tau(0)} \xi(X_k)\bigg) = 0 \quad \text{and} \quad \tilde{E}_{\nu^*} f^2\bigg(\sum_{k=0}^{\tau(0)} \xi(X_k)\bigg) < +\infty \quad f \in B^*$$

and

$$(3.17) \qquad \sigma_f^2 = \pi^*(\alpha^*) \cdot \tilde{E}_{\nu^*} f^2\bigg(\sum_{k=0}^{\tau(0)} \xi(X_k)\bigg) \quad f \in B^*.$$

Let

$$(3.18) \quad S(f) = \pi^*(\alpha^*) \cdot \tilde{E}_{\nu^*}\bigg\{\bigg(\sum_{k=0}^{\tau(0)} \xi(X_k)\bigg) f\bigg(\sum_{k=0}^{\tau(0)} \xi(X_k)\bigg)\bigg\} \quad f \in B^*,$$

where the expectation is defined as a Pettis integral. Then, S defines a continuous linear operator from B^* to B (see Lemma 2.1 in [26]). Define

$$(3.19) \qquad ||S(f)||_H = \pi^*(\alpha^*)^{1/2} \cdot \bigg\{\tilde{E}_{\nu^*} f^2\bigg(\sum_{k=0}^{\tau(0)} \xi(X_k)\bigg)\bigg\}^{1/2} \quad f \in B^*.$$

By Lemma 2.1 in [26], the completion H of $Im(S)$ w.r.t. the norm $||\cdot||_H$ is a subspace of B. Then $(H, ||\cdot||_H)$ is the reproducing kernel Hilbert space generated by $\{\sigma_f^2; \; f \in B^*\}$ with the covariance operator S given in (3.18).

2. In the general case, we have

$$(3.20) \qquad P^m \geq bI_C \otimes \nu$$

for some small set C, $m \geq 1$, $b > 0$ and some probability measure ν on $(E, \mathcal{E})$. Consider the split chain $\tilde{P}$ (possibly in the general sense) generated by the minorization (3.20). Let the regeneration times $\{\tau(k)\}_{k \geq 0}$ be given in (I-2.19). By Corollary I-2.4, the sequence $\{\xi_k\}_{k \geq 1}$ of B-valued random variables given by

$$(3.21) \qquad \xi_k = \sum_{j=m(\tau(k-1)+1)}^{m\tau(k)+m-1} \xi(X_j) \qquad k = 1, 2, \cdots$$

is 1-dependent with the common distribution

$$\tilde{\mathcal{L}}_{\tilde{P}_{\nu^*}}\left(\sum_{j=0}^{m\tau(0)+m-1} \xi(X_j)\right).$$

Assume that

$$(3.22) \qquad \int (f \circ \xi)^2(x) \pi(dx) < +\infty \qquad \forall f \in B^*.$$

By Theorem II-2.3 we have

$$(3.23) \qquad Ef(\xi_1) = 0 \quad \text{and} \quad Ef^2(\xi_1) < +\infty \qquad \forall f \in B^*,$$

$$(3.24) \qquad \sigma_f^2 = m^{-1}\pi^*(\alpha^*)\{Ef^2(\xi_1) + 2Ef(\xi_1)f(\xi_2)\} \qquad \forall f \in B^*.$$

Define

$$(3.25) \quad S(f) = m^{-1}\pi^*(\alpha^*)\{E\xi_1 f(\xi_1) + E\xi_1 f(\xi_2) + Ef(\xi_1)\xi_2\} \qquad f \in B^*.$$

Then, S defines a continuous linear operator from B^* to B. Let the inner product $(\cdot, \cdot)_H$ on $Im(S)$ be defined by

$$(3.26) \quad \begin{aligned}(Sf, Sg)_H = m^{-1}\pi^*(\alpha^*)\{&Ef(\xi_1)g(\xi_1) + Ef(\xi_1)g(\xi_2) \\ &+ Eg(\xi_1)f(\xi_2)ig\} \qquad f, g \in B^*\end{aligned}$$

and let $H \subset B$ be the completion of of $Im(S)$ w.r.t. the norm $\|\cdot\|_H$ (see Lemma 1 in [14]). Then $(H, \|\cdot\|_H)$ is the reproducing kernel Hilbert space generated by $\{\sigma_f^2;\ f \in B^*\}$ with the covariance operator S given in (3.25).

III-4. The LIL in vector setting

To answer the question raised at the end of Section 2, we start with a careful examination of the conditions needed for the LIL. Suppose the LIL is satisfied. By Orey's theorem (Theorem 2.6, p167, [40]) there exists a constant $\Gamma \geq 0$ such that

$$(4.1) \qquad \limsup_{n\to\infty} \frac{\|S_n\|}{\sqrt{2nL_2 n}} = \Gamma \quad \text{a.s.}$$

Then clearly, we have

$$\left\{S_n/\sqrt{2nL_2 n}\right\}_{n\geq 1} \text{ is bounded in probability.}$$

To get a more precise form of the LIL, we shall also consider a slightly stronger condition:

$$S_n/\sqrt{2nL_2 n} \longrightarrow 0 \quad \text{in probability.}$$

Assumptions of this type have become standard in the context of almost sure limit theorems for independent infinite dimensional random variables and in general, they can not be derived from the other conditions given for the LIL. We will see that they also apply very well in the Markovian context.

Besides, one can easily see that (4.1) implies that for each $f \in B^*$,

$$\limsup_{n\to\infty} \frac{f(S_n)}{\sqrt{2nL_2 n}} < +\infty \quad \text{a.s.}$$

By Theorem 2.2 (with $f \circ \xi$ instead of ξ), we must have $\xi \in WCLT$ and (where σ_f^2 is the g.c.f. associated with ξ)

$$(4.2) \qquad \limsup_{n \to \infty} \frac{f(S_n)}{\sqrt{2nL_2n}} = \sigma_f \quad \text{a.s.} \quad \forall f \in B^*,$$

provided

$$(4.3) \qquad \int (f \circ \xi)^2(x) \pi(dx) < +\infty \quad \forall f \in B^*.$$

The relation (4.2) suggests that in describing the cluster set phenomena of the LIL, the g.c.f. $\{\sigma_f^2;\ f \in B^*\}$ might play the same role as the covariance function does in the i.i.d. context (see Chaper 8 in Ledoux-Talagrand (1991)). Recall that given $\xi \in WCLT$ with the g.c.f. $\{\sigma_f^2;\ f \in B^*\}$, $(H, ||\cdot||_H)$ denotes, whenever it exists, the reproducing kernel Hilbert space generated by $\{\sigma_f^2;\ f \in B^*\}$ and K the closed unit ball of $(H, ||\cdot||_H)$. According to Lemma 3.1-(i),

$$(4.4) \qquad \sigma \equiv \sup_{f \in B_1^*} \sigma_f < +\infty.$$

For a sequence $\{x_n\}_{n \geq 1}$ in B, we write $C(\{x_n\}_{n \geq 1})$ for the cluster set of $\{x_n\}_{n \geq 1}$. We define

$$d(x, F) = \inf_{y \in F} ||x - y||$$

for given $x \in B$ and $F \subset B$. The argument used in the proof of Lemma 1 in Kuelbs (1981), along with Orey' theorem (Theorem 2.6, p167, [40]) instead of the Hewitt-Savage 0-1 law, implies there exists a non-random set $A \subset B$ such that

$$(4.5) \qquad C\left(\{S_n/\sqrt{2nL_2n}\}_{n \geq 1}\right) = A \quad \text{a.s.}$$

Identified with A, $C\left(\{S_n/\sqrt{2nL_2n}\}_{n \geq 1}\right)$ is thus viewed as a non-random subset of B.

Notice that (4.3) implies the existence of $(H, ||\cdot||_H)$ via the dicussion following to Lemma 3.1. Like in the i.i.d. context, we have the following:

Proposition 4.1. *Suppose (4.3) holds and*

(4.6) $$\limsup_{n\to\infty} \frac{||S_n||}{\sqrt{2nL_2n}} < +\infty \quad a.s.$$

Then:

(i) $\xi \in WCLT$ *and*

(4.7) $$\limsup_{n\to\infty} \frac{||S_n||}{\sqrt{2nL_2n}} \geq \sigma \quad a.s.,$$

(4.8) $$C\left(\{S_n/\sqrt{2nL_2n}\}_{n\geq 1}\right) \subset K.$$

(ii) The following two statements (a) and (b) are equivalent:

(a) $\{S_n/\sqrt{2nL_2n}\}_{n\geq 1}$ *is almost surely relatively compact in B;*

(b) $$\begin{cases} K \text{ is compact in } B, \\ \lim_{n\to\infty} d(S_n/\sqrt{2nL_2n}, K) = 0 \quad a.s., \\ C(\{S_n/\sqrt{2nL_2n}\}) = K, \\ \limsup_{n\to\infty} ||S_n||/\sqrt{2nL_2n} = \sigma \quad a.s. \end{cases}$$

Proof. The assertion $\xi \in WCLT$ and (4.7) follow from Theorem 2.2 and the rest from (3.9), (4.2), (4.5) and the argument used in the proof of Proposition 2.2 in Chen ([14]). ∎

Clearly, (4.6) and the statement (a) in (ii) of Proposition 4.1 are the same if B is a finite dimensional space, in which case the limiting behavior of the LIL is, if satisfied, basically determined by K. However, this is not the case for the general infinite dimensional space B even in the i.i.d. context (see Chapter 8 in Ledoux-Talagrand (1991)). In the literature the LIL described by (4.6) is called the **bounded** LIL and the LIL described by the equivalent statements in (ii) of Proposition 4.1 is called the **compact** LIL.

To have the LIL, we also need some integrability condition. Let $\mathcal{S}$ be the class of small sets in $(E, \mathcal{E})$ and define

$$(4.9) \quad \mathcal{S}_{LIL}(\xi) = \left\{ C \in \mathcal{S}; \; \int_C \pi(dx) E_x \max_{n \leq \tau_C} \left(||S_n||^2 / L_2 ||S_n|| \right) < +\infty \right\}.$$

By Theorem I-4.1,

$$(4.10) \quad \mathcal{S}_{LIL}(\xi) = \phi \; \text{ or } \; \mathcal{S}.$$

The following lemma states that one of the necessary conditions for the LIL is that $\mathcal{S}_{LIL}(\xi) = \mathcal{S}$.

Lemma 4.2. *If (4.6) holds then,* $\mathcal{S}_{LIL}(\xi) = \mathcal{S}$*, i.e.,*

$$(4.11) \quad \int_C \pi(dx) E_x \max_{n \leq \tau_C} \left(||S_n||^2 / L_2 ||S_n|| \right) < +\infty$$

hold for all small sets C.

Proof. By (4.10), we need only to prove (4.11) for some small set C. We choose C to be a sub-π small set and we thus have the following minorization

$$(4.12) \quad P^m \geq b I_C \otimes \pi_C.$$

Define the stopping times $\{\tau_C(k)\}_{k \geq 1}$ and $\{\tau'_C(k)\}_{k \geq 1}$ as follows:

$$\begin{cases} \tau_C(1) = \tau_C, \\ \tau_C(k+1) = \inf\{n > \tau_C(k); \; X_n \in C\} \quad (k \geq 1), \end{cases}$$

$$\begin{cases} \tau'_C(1) = \inf\{n \geq m+1;\ X_n \in C\}, \\ \tau'_C(k+1) = \inf\{n > m + \tau'_C(k);\ X_n \in C\} \quad (k \geq 1). \end{cases}$$

By the law of the large numbers for Markov chains (Theorem 17.1.7 in Meyn-Tweedie (1993)),

$$\lim_{n \to \infty} \frac{1}{n} \sum_{k=1}^{n} I_C(X_k) = \pi(C) \quad \text{a.s.}$$

Note that, by definition, for each $l \geq 1$

$$\left\{ \sum_{k=1}^{n} I_C(X_k) \geq l \right\} = \{\tau_C(l) \leq n\}.$$

Hence, one can conclude from this relation that

$$\lim_{n \to \infty} \frac{1}{n} \tau_C(n) = \pi(C)^{-1} \quad \text{a.s.}$$

On the other hand, by definition, one can see that $\tau'_C(n) \leq \tau_C((m+1)n)$ a.s. for all $n \geq 1$. Consequently,

$$\limsup_{n \to \infty} \frac{1}{n} \tau'_C(n) \leq (m+1)\pi(C)^{-1} \quad \text{a.s.}$$

By Orey's theorem (Theorem 2.6, p167, [40]), (4.6) implies (4.1) for some constant $\Gamma \geq 0$. Equivalently,

$$\limsup_{n \to \infty} \max_{k \leq n} \|S_k\| / \sqrt{2nL_2 n} = \Gamma \quad \text{a.s.}$$

In particular,

$$\limsup_{n \to \infty} \max_{k \leq m + \tau'_C(n)} \|S_k\| / \sqrt{2(m + \tau'_C(n)) L_2(m + \tau'_C(n))} \leq \Gamma \quad \text{a.s.,}$$

$$\limsup_{n \to \infty} \max_{k \leq \tau'_C(n+1)} \|S_k\| / \sqrt{2\tau'_C(n+1) L_2 \tau'_C(n+1)} \leq \Gamma \quad \text{a.s.}$$

Therefore,

$$\limsup_{n\to\infty} \max_{k\leq m+\tau'_C(n)} \|S_k\|/\sqrt{2nL_2 n} \leq (m+1)^{1/2}\pi(C)^{-1/2}\Gamma \quad \text{a.s.,}$$

$$\limsup_{n\to\infty} \max_{k\leq \tau'_C(n+1)} \|S_k\|/\sqrt{2nL_2 n} \leq (m+1)^{1/2}\pi(C)^{-1/2}\Gamma \quad \text{a.s.}$$

Let

$$\eta_n = \max_{m+\tau'_C(n)\leq k\leq \tau'_C(n+1)} \|S_k - S_{m+\tau'_C(n)}\| \quad n=1,2,\cdots.$$

By the triangle inequality, there exists a constant $M > 0$ such that

(4.13) $$\limsup_{n\to\infty} \eta_n/\sqrt{2nL_2 n} < M \quad \text{a.s.}$$

On the other hand, for each $n \geq 1$ we have

$$\eta_n = \left(\max_{m\leq k\leq \tau'_C(1)} \|S_k - S_m\|\right) \circ \theta^{\tau'_C(n)} \quad \text{a.s.}$$

Let the integers $N \geq 1$ and $l \geq 2$ be fixed but arbitrary. Given an initial distribution μ, using the Markov property gives

$$P_\mu\Big(\bigcap_{n=N+1}^{N+l} \{\eta_n \leq M\sqrt{2nL_2 n}\}\Big)$$

$$= E_\mu\Big(\prod_{n=N+1}^{N+l-1} I_{\{\eta_n \leq M\sqrt{2nL_2 n}\}}$$

$$\times P_{X_{\tau'_C(N+l)}}\Big\{\max_{m\leq k\leq \tau'_C(1)} \|S_k - S_m\| \leq M\sqrt{2(N+l)L_2(N+l)}\Big\}\Big).$$

Note that

$$\max_{m\leq k\leq \tau'_C(1)} \|S_k - S_m\| = \Big(\max_{k\leq \tau_C} \|S_k\|\Big) \circ \theta^m \quad \text{a.s.}$$

By the Markov property and (4.12) for each $x \in C$,

$$P_x\Big\{\max_{m\leq k\leq \tau'_C(1)} \|S_k - S_m\| > M\sqrt{2(N+l)L_2(N+l)}\Big\}$$

$$= \int P^m(x,dx_1) P_{x_1}\Big\{\max_{k\leq \tau_C} \|S_k\| > M\sqrt{2(N+l)L_2(N+l)}\Big\}$$

$$\geq b P_{\pi_C}\Big\{\max_{k\leq \tau_C} \|S_k\| > M\sqrt{2(N+l)L_2(N+l)}\Big\}.$$

Consequently,

$$P_{X_{\tau'_C(N+l)}}\{\max_{m\leq k\leq \tau'_C(1)}\|S_k - S_m\| \leq M\sqrt{2(N+l)L_2(N+l)}\}$$
$$\leq 1 - bP_{\pi_C}\{\max_{k\leq \tau_C}\|S_k\| > M\sqrt{2(N+l)L_2(N+l)}\} \quad \text{a.s.}$$

Therefore,

$$P_\mu\Big(\bigcap_{n=N+1}^{N+l}\{\eta_n \leq M\sqrt{2nL_2n}\}\Big)$$
$$\leq P_\mu\Big(\bigcap_{n=N+1}^{N+l-1}\{\eta_n \leq M\sqrt{2nL_2n}\}\Big)$$
$$\times \Big(1 - bP_{\pi_C}\{\max_{k\leq \tau_C}\|S_k\| > M\sqrt{2(N+l)L_2(N+l)}\}\Big).$$

Repeating this procedure gives

$$P_\mu\Big(\bigcap_{n=N+1}^{N+l}\{\eta_n \leq M\sqrt{2nL_2n}\}\Big)$$
$$\leq \prod_{n=N+1}^{N+l}\Big(1 - bP_{\pi_C}\{\max_{k\leq \tau_C}\|S_k\| > M\sqrt{2nL_2n}\}\Big).$$

Letting $l \longrightarrow +\infty$ gives that for each $N \geq 1$,

$$P_\mu\Big(\bigcap_{n=N+1}^{\infty}\{\eta_n \leq M\sqrt{2nL_2n}\}\Big)$$
$$\leq \prod_{n=N+1}^{\infty}\Big(1 - bP_{\pi_C}\{\max_{k\leq \tau_C}\|S_k\| > M\sqrt{2nL_2n}\}\Big).$$

Since (4.13) implies that

$$P_\mu\Big(\bigcap_{n=N+1}^{\infty}\{\eta_n \leq M\sqrt{2nL_2n}\}\Big) \longrightarrow 1 \quad (N \to \infty)$$

we must have

$$\sum_{n=1}^{\infty} P_{\pi_C}\{\max_{k\leq \tau_C}\|S_k\| > M\sqrt{2nL_2n}\} < +\infty,$$

which implies (4.11). ∎

We now state the main result in this section.

Theorem 4.3. *Let $\{X_n\}_{n\geq 0}$ be an ergodic Markov chain and assume (4.3) holds. Then, in order that the bounded LIL hold, i.e.,*

$$(4.14) \qquad \limsup_{n\to\infty} \frac{\|S_n\|}{\sqrt{2nL_2 n}} < +\infty \quad a.s.,$$

it is sufficient and necessary that the following three conditions are fulfilled:

(i) $\qquad \xi \in WCLT;$

(ii) for some (all) small set C,

$$\int_C \pi(dx) E_x \max_{n\leq \tau_C} \left(\|S_n\|^2/L_2\|S_n\|\right) < +\infty;$$

(iii) $\qquad \{S_n/\sqrt{2nL_2 n}\}_{n\geq 1}$ *is bounded in probability.*

Further,

$$(4.15) \qquad \limsup_{n\to\infty} \frac{\|S_n\|}{\sqrt{2nL_2 n}} = \sigma \quad a.s.,$$

$$(4.16) \qquad \lim_{n\to\infty} d(S_n/\sqrt{2nL_2 n}, K) = 0 \quad a.s.,$$

and

$$(4.17) \qquad C\left(\{S_n/\sqrt{2nL_2 n}\}_{n\geq 1}\right) = K \quad a.s.$$

whenever (i), (ii) hold and (iii) is strengthened into

(iii') $\qquad S_n/\sqrt{2nL_2 n} \longrightarrow 0$ *in probability.*

Let us make some remarks before the proof. First, the condition (i) can be verified by using the results in Chapter II. By Theorem II-3.1 (with $f \circ \xi$

instead of ξ for each $f \in B^*$), for example, we have $\xi \in WCLT$ with the g.c.f. $\{\sigma_f^2;\ f \in B^*\}$ given by

$$(4.18) \quad \sigma_f^2 = \int (f \circ \xi)^2(x)\pi(dx) + 2\int \sum_{n=1}^{\infty} f \circ \xi(x) P^n(f \circ \xi)(x)\pi(dx) \quad f \in B^*$$

under (4.3) and the assumption that

$$(4.19) \quad \sum_{n=1}^{\infty} f \circ \xi(\cdot) P^n(f \circ \xi)(\cdot) \quad \text{converges in } L^1(\pi) \text{ for each } f \in B^*.$$

Second, we will see from Remark 2.3 and the following proof that when $\{X_n\}_{n \geq 0}$ has a small set of order 1, the conclusion of Theorem 4.3 stands even without (4.3).

Proof of Theorem 4.3. The necessity of the conditions (i), (ii) and (iii) for the bounded LIL (4.14) is proven in the previous discussion. We now prove (4.14) under (i), (ii) and (iii). Let the small set C be fixed. By definition there are some $m \geq 1$, $0 < b < 1/2$ and some probability measure ν on $(E, \mathcal{E})$ such that

$$(4.20) \quad P^m \geq 2b I_C \otimes \nu.$$

We may assume that there exists a constant $\lambda > 0$ such that

$$(4.21) \quad C \subset \left\{ x;\ P_x\left\{ \sum_{j=1}^{m-1} \|\xi(X_j)\| > \lambda \right\} \leq \frac{b}{4} \right\},$$

for otherwise we can take $\lambda > 0$ sufficiently large so that

$$D \equiv C \cap \left\{ x;\ P_x\left\{ \sum_{j=1}^{m-1} \|\xi(X_j)\| > \lambda \right\} \leq \frac{b}{4} \right\} \in \mathcal{E}^+$$

and use the small set D instead of C.

Consider the split chain $\tilde{P}$ (possibly in the general sense) generated by the minorization

(4.22) $$P^m \geq bI_C \otimes \nu.$$

Let the regeneration times $\{\tau(k)\}_{k\geq 0}$ be defined as in (I-2.19). By Theorem I-4.2,

(4.23) $$\tilde{E}_{\nu^*} \max_{n \leq m(\tau(0)+1)} \left(\|S_n\|^2/L_2\|S_n\|\right) < +\infty.$$

For each $n \geq 1$, let $i(n)$ be given as in (I-3.10). One can see that for each $n \geq 1$

(4.24) $$S_n = S_{m(\tau(0)+1)\wedge n} + \sum_{k=1}^{i(n)-1} \xi_k + \sum_{j=m(\tau((i(n)-1)\vee 0)+1)}^{n-1} \xi(X_j) \quad \text{a.s.},$$

where the 1-dependent B-valued random sequence $\{\xi_k\}_{k\geq 1}$ is defined as in (3.21). One can also see that on the event $\{\tau(0) \leq [n/m] - 1\}$,

(4.25) $$m(\tau(i(n)-1)+1) \leq n-1 \leq m(\tau(i(n)-1)+1) \quad \text{a.s.}$$

To prove (4.14), we need to control each of the three terms on the right-hand side of (4.24). For the first term, since

$$\|S_{m(\tau(0)+1)\wedge n}\| \leq \sum_{j=0}^{m\tau(0)+m-1} \|\xi(X_j)\| \quad \text{a.s.}$$

we have

(4.26) $$\lim_{n\to\infty} \|S_{m(\tau(0)+1)\wedge n}\|/\sqrt{2nL_2n} = 0 \quad \text{a.s.}$$

We now control the third term. More precisely, we prove that

(4.27) $$\lim_{n\to\infty} \left\|\sum_{j=m(\tau((i(n)-1)\vee 0)+1)}^{n-1} \xi(X_j)\right\|/\sqrt{2nL_2n} = 0 \quad \text{a.s.}$$

Consider the following 1-dependent random variables:

$$\max_{m(\tau(n-1)+1)\leq k<m(\tau(n)+1)} \left\|\sum_{j=m(\tau(n-1)+1)}^{k} \xi(X_j)\right\| \quad n=1,2,\cdots.$$

By Corollary I-2.4, they have the common distribution:

$$\tilde{\mathcal{L}}_{\tilde{P}_{\nu^*}}\left(\max_{k\leq m(\tau(0)+1)} \|S_k\|\right).$$

Hence, (4.23) implies that for any $\epsilon > 0$,

$$\sum_n P\left\{\max_{m(\tau(n-1)+1)\leq k<m(\tau(n)+1)} \left\|\sum_{j=m(\tau(n-1)+1)}^{k} \xi(X_j)\right\| \geq \epsilon\sqrt{2nL_2n}\right\}$$
$$< +\infty.$$

By the Borel-Cantelli lemma,

$$\lim_{n\to\infty} \max_{m(\tau(n-1)+1)\leq k<m(\tau(n)+1)} \left\|\sum_{j=m(\tau(n-1)+1)}^{k} \xi(X_j)\right\|/\sqrt{2nL_2n}$$
$$= 0 \quad \text{a.s.}$$

In particular,

$$\lim_{n\to\infty} \max_{m(\tau(i(n)-1)+1)\leq k<m(\tau(i(n))+1)} \left\|\sum_{j=m(\tau(i(n)-1)+1)}^{k} \xi(X_j)\right\|/\sqrt{2i(n)L_2i(n)}$$
$$= 0 \quad \text{a.s.}$$

By the law of the large numbers for Markov chains (Theorem 17.1.7 in Meyn-Tweedie (1993)),

(4.28) $$\lim_{n\to\infty} \frac{i(n)}{n} = m^{-1}\pi^*(\alpha^*) \quad \text{a.s.}$$

Therefore,

$$\lim_{n\to\infty} \max_{m(\tau(i(n)-1)+1)\leq k<m(\tau(i(n))+1)} \left\|\sum_{j=m(\tau(i(n)-1)+1)}^{k} \xi(X_j)\right\|/\sqrt{2nL_2n}$$
$$= 0 \quad \text{a.s.}$$

From (4.25), on the event $\{\tau(0) \leq [n/m] - 1\}$,

$$\Big\| \sum_{j=m(\tau((i(n)-1)\vee 0)+1)}^{n-1} \xi(X_j) \Big\|$$
$$\leq \max_{m(\tau(i(n)-1)+1)\leq k<m(\tau(i(n))+1)} \Big\| \sum_{j=m(\tau(i(n)-1)+1)}^{k} \xi(X_j) \Big\| \quad \text{a.s.}$$

Hence we have (4.27).

Recall (Theorem 1 in Chen ([14])) that if the 1-dependent B-valued random sequence $\{\xi_k\}_{k\geq 1}$ satisfies

(4.29) $\qquad Ef(\xi_1) = 0 \quad \text{and} \quad Ef^2(\xi_1) < +\infty \quad \forall f \in B^*,$

(4.30) $\qquad E\big(\|\xi_1\|^2/L_2\|\xi_1\|\big) < +\infty,$

and

(4.31) $\qquad \Big\{\sum_{k=1}^{n} \xi_k / \sqrt{2nL_2n}\Big\}_{n\geq 1}$ is bounded in probability,

then

(4.32) $\qquad \limsup_{n\to\infty} \Big\|\sum_{k=1}^{n} \xi_k\Big\| / \sqrt{2nL_2n} < +\infty \quad \text{a.s.}$

Further,

(4.33) $\qquad \limsup_{n\to\infty} \Big\|\sum_{k=1}^{n} \xi_k\Big\| / \sqrt{2nL_2n} = \sigma' \quad \text{a.s.},$

(4.34) $\qquad \lim_{n\to\infty} d\Big(\sum_{k=1}^{n} \xi_k / \sqrt{2nL_2n}, K'\Big) = 0 \quad \text{a.s.},$

and

(4.35) $$C\left(\left\{\sum_{k=1}^{n}\xi_k/\sqrt{2nL_2n}\right\}_{n\geq 1}\right) = K' \quad \text{a.s.}$$

whenever (4.29), (4.30) hold and (4.31) is strengthened into

(4.36) $$\sum_{k=1}^{n}\xi_k/\sqrt{2nL_2n} \longrightarrow 0 \quad \text{in probability.}$$

Here K' is the closed unit ball of the reproducing kernel Hilbert space generated by (see Lemma 1 of [14] for the details) the g.c.f. $\{\sigma'^2_f;\ f \in B^*\}$ given by

(4.37) $$\sigma'^2_f = Ef^2(\xi_1) + 2Ef(\xi_1)f(\xi_2) \quad f \in B^*$$

and

$$\sigma' = \sup_{f \in B^*_1} \sigma'_f.$$

By Theorem II-2.3, (4.29) is satisfied and

(4.38) $$\sigma^2_f = m^{-1}\pi^*(\alpha^*)\sigma'^2_f \quad f \in B^*.$$

By Corollary I-2.4, the 1-dependent random variables $\{\xi_k\}_{k\geq 1}$ have the common distribution

$$\mathcal{L}_{\tilde{P}_{\nu^*}}\left(\sum_{j=0}^{m\tau(0)+m-1}\xi(X_j)\right) = \mathcal{L}_{\tilde{P}_{\nu^*}}\left(S_{m(\tau(0)+1)}\right).$$

Hence (4.30) follows from (4.23). We now verify (4.31). Due to Theorem I-3.1, we need only to prove it in the case when ν^* is the initial distribution. By the condition (iii), for sufficiently large $M > 0$,

$$\sup_n P_\nu\{\|S_n\| \geq M\sqrt{2nL_2n}\} < \frac{1}{4}.$$

In view of (4.21), by the maximal inequality (I-3.12) in Theorem I-3.4 we have that for fixed $0 < \lambda < \pi^*(\alpha^*)$ and sufficiently large n,

$$\tilde{P}_{\nu^*}\left\{||S_{m\tau([n\lambda/m])+1}|| \geq 3M\sqrt{2nL_2 n}, \ i(n) \geq \left[\frac{n\lambda}{m}\right]+1\right\}$$
$$\leq \tilde{P}_{\nu^*}\left\{\max_{0 \leq k < i(n)} ||S_{m\tau(k)+1}|| \geq 3M\sqrt{2nL_2 n}\right\}$$
$$\leq 2P_{\nu}\{||S_n|| \geq M\sqrt{2nL_2 n}\}.$$

From (4.28) we have

$$\tilde{P}_{\nu^*}\left\{i(n) \geq \left[\frac{n\lambda}{m}\right]+1\right\} \longrightarrow 1 \quad (n \to \infty).$$

Therefore, it follows from the condition (iii) that

$$\left\{S_{m\tau([n\lambda/m])+1}/\sqrt{2nL_2 n}\right\}_{n \geq 1} \text{ is bounded in probability.}$$

Or, equivalently,

$$\left\{S_{m\tau(n)+1}/\sqrt{2nL_2 n}\right\}_{n \geq 1} \text{ is bounded in probability.}$$

Note that

$$S_{m\tau(n)+1} = S_{m(\tau(0)+1) \wedge n} + \sum_{k=1}^{n} \xi_k - \sum_{j=m\tau(n)+1}^{m\tau(n)+m-1} \xi(X_j) \quad \text{a.s.,}$$

and

$$\tilde{P}_{\nu^*}\left\{\left\|\sum_{j=m\tau(n)+1}^{m\tau(n)+m-1} \xi(X_j)\right\| \geq t\right\} = \tilde{E}_{\nu^*}\left(\tilde{P}_{\Phi_{\tau(n)}}\left\{\left\|\sum_{j=1}^{m-1} \xi(X_j)\right\| \geq t\right\}\right)$$
$$= \tilde{E}_{\alpha^*}\left(\tilde{P}_{\Phi_{\tau_{\alpha^*}(n)}}\left\{\left\|\sum_{j=1}^{m-1} \xi(X_j)\right\| \geq t\right\}\right)$$
$$= \pi^*(\alpha^*)^{-1} \int_{\alpha^*} \pi^*(d(x,y)) \tilde{P}_{(x,y)}\left\{\left\|\sum_{j=1}^{m-1} \xi(X_j)\right\| \geq t\right\} \longrightarrow 0$$

as $t \to +\infty$, where

$$\begin{cases} \tau_{\alpha^*}(0) = 0 \text{ and } \tau_{\alpha^*}(1) = \inf\{n \geq 1; \ \Phi_n \in \alpha^*\} \\ \tau_{\alpha^*}(k+1) = \inf\{n > \tau_{\alpha^*}(k); \ X_n \in \alpha^*\} \qquad (k \geq 1). \end{cases}$$

The first and the second equalities above follow from the Markov property, and the third from the fact that $\pi^*(\alpha^*)^{-1}\pi^*(\alpha^* \cap \cdot)$ is the invariant distribution of the Markov chain $\{\Phi_{\tau_{\alpha^*}(k)}\}_{k \geq 0}$. In view of (4.26) we have (4.31).

¿From (4.28) and (4.32) we have

$$(4.39) \qquad \limsup_{n \to \infty} \left\| \sum_{k=1}^{i(n)-1} \xi_k \right\| / \sqrt{2nL_2n} < +\infty \quad \text{a.s.}$$

Therefore, (4.14) follows from (4.24), (4.26), (4.27) and (4.29).

Further, suppose the condition (iii) can be strengthened into (iii'). Instead of (4.31), we can verify (4.36) in a similar way. Hence we have (4.33), (4.34) and (4.35). By (4.28), (4.33) and (4.34) imply that

$$(4.40) \qquad \limsup_{n \to \infty} \left\| \sum_{k=1}^{i(n)-1} \xi_k \right\| / \sqrt{2nL_2n} \leq m^{-1/2} \pi^*(\alpha^*)^{1/2} \sigma' \quad \text{a.s.,}$$

and

$$(4.41) \qquad \lim_{n \to \infty} d\left(\sum_{k=1}^{i(n)-1} \xi_k / \sqrt{2nL_2n}, \sqrt{\frac{\pi^*(\alpha^*)}{m}} K' \right) = 0 \quad \text{a.s.,}$$

respectively. Since,

$$S_{m(\tau(n)+1)} = S_{m(\tau(0)+1)} + \sum_{k=1}^{n} \xi_k \quad \text{a.s.,}$$

from (4.35) we have

$$C\left(\left\{S_{m(\tau(n)+1)} / \sqrt{2nL_2n}\right\}_{n \geq 1}\right) = K' \quad \text{a.s.}$$

By the law of the large numbers,

$$\lim_{n\to\infty} \frac{\tau(n)}{n} = \tilde{E}_{\nu^*}\tau(0) = \frac{1}{\pi^*(\alpha^*)} \quad \text{a.s.}$$

Consequently,

(4.42)
$$C\left(\left\{S_{m(\tau(n)+1)}/\sqrt{2(m(\tau(n)+1))L_2(m(\tau(n)+1))}\right\}_{n\geq 1}\right)$$
$$= \sqrt{\frac{\pi^*(\alpha^*)}{m}} K' \quad \text{a.s.}$$

From (4.38) and the construction of the reproducing kernel Hilbert space, one can see that $\sigma = m^{-1/2}\pi^*(\alpha^*)^{1/2}\sigma'$ and $K = m^{-1/2}\pi^*(\alpha^*)^{1/2}K'$. In view of (4.24), (4.26) and (4.27), assertion (4.16) follows from (4.41) and, by (4.40),

$$\limsup_{n\to\infty} \frac{\|S_n\|}{\sqrt{2nL_2n}} \leq \sigma \quad \text{a.s.}$$

This, together with (4.7) in Proposition 4.1, implies (4.15). From (4.42) we have

$$C\left(\{S_n/\sqrt{2nL_2n}\}_{n\geq 1}\right) \supset K \quad \text{a.s.}$$

Comparing this with (4.8) in Proposition 4.1 we have (4.17). ∎

III-5. Some consequences

The first part of Theorem 4.3 basically gives sufficient and necessary conditions for the bounded LIL. Using the second part, we have the following characterization of the compact LIL.

Theorem 5.1. *Let $\{X_n\}_{n\geq 0}$ be an ergodic Markov chain and assume (4.3) holds. Then, in order that the compact LIL hold, i.e., the equivalent statements in (ii) of Proposition 4.1 hold, it is sufficient and necessary that the following three conditions are fulfilled:*

(i) $\xi \in WCLT$ with the g.c.f. $\{\sigma_f^2;\ f \in B^*\}$ weak-star sequentially continuous,

(ii) for some (all) small set C,

$$\int_C \pi(dx) E_x \max_{n \leq \tau_C} \left(||S_n||^2/L_2||S_n||\right) < +\infty,$$

(iii) $\qquad\qquad S_n/\sqrt{2nL_2n} \longrightarrow 0$ in probability.

Proof. Although the situations are different, the argument we use here is standard in the independent context. For the sake of completeness, we include it.

First we prove the necessary part. By Theorem 4.3, $\xi \in WCLT$ and (ii) holds. According to Proposition 4.1, K is compact. Equivalently (Lemma 3.1-(iii)), $\{\sigma_f^2;\ f \in B^*\}$ is weak-star sequentially continuous. Clearly, the compact LIL implies the uniform tightness of $\{\mathcal{L}(S_n/\sqrt{2nL_2n})\}_{n\geq 1}$: Given $\epsilon > 0$ and the initial distribution μ there is a compact subset $K_{\epsilon,\mu}$ of B such that

$$\inf_{n\geq 1} P_\mu\{S_n/\sqrt{2nL_2n} \in K_{\epsilon,\mu}\} \geq 1 - \epsilon.$$

Note that $\xi \in WCLT$ implies that for each $f \in B^*$,

$$f\left(S_n/\sqrt{2nL_2n}\right) \longrightarrow 0 \text{ in probability}.$$

This, together with the uniform tightness of $\{\mathcal{L}(S_n/\sqrt{2nL_2n})\}_{n\geq 1}$, implies the condition (iii).

The idea for the sufficient part is the finite dimensional approximation. Let $\{f_k\}_{k\geq 1}$ be a weak star dense subset of B_1^* and let S be the covariance operator associated with the g.c.f. $\{\sigma_f^2;\ f \in B^*\}$. By the density of $Im(S)$ in H, one can obtain a sequence $\{g_k\}_{k\geq 1}$ in B^* from $\{f_k\}_{k\geq 1}$ by the usual

Gram-Schmidt orthogonalization method such that $\{S(g_k)\}_{k\geq 1}$ is a complete orthonormal basis in H. Then, the linear operators

$$\Pi_N(x) = \sum_{k=1}^{N} g_k(x) S(g_k) \quad \text{and} \quad Q_N(x) = x - \Pi_N(x) \quad (N \geq 1)$$

are continuous from B into B. When restricted to H, Π_N and Q_N are orthogonal projections onto their ranges. In particular, $\|Q_N(x)\|_H \longrightarrow 0$ for each $x \in H$ as $N \to \infty$. From (3.6), one can see how this implies

(5.1) $$\lim_{N \to \infty} \|Q_N(x)\| = 0 \quad x \in H.$$

For each $N \geq 1$, by (4.16) in Theorem 4.3 we have

$$\limsup_{n \to \infty} \|Q_N(S_n/\sqrt{2nL_2n})\| \leq \sup_{x \in K} \|Q_N(x)\| \quad \text{a.s.}$$

To prove the compact LIL, therefore, one needs only to show

(5.2) $$\lim_{N \to \infty} \sup_{x \in K} \|Q_N(x)\| = 0.$$

To do this, we first claim that for each $N \geq 1$,

(5.3) $$Q_N(K) \supset Q_{N+1}(K).$$

Indeed, for given $x \in Q_{N+1}(K)$ one has that $Q_{N+1}(x) \in K$. Hence,

$$Q_{N+1}(x) = Q_N(Q_{N+1}(x)) \in Q_N(K).$$

We then prove

(5.4) $$\bigcap_{N=1}^{\infty} Q_N(K) = \{0\}.$$

In fact, if $x \in \bigcap_{N=1}^{\infty} Q_N(K)$ then for each $N \geq 1$, there exists $x_N \in K$ such that $x = Q_N(x_N)$. Therefore,

$$Q_N(x) = Q_N(x_N) \quad N = 1, 2, \cdots.$$

This, along with (5.1), implies

$$x = \lim_{N \to \infty} Q_N(x_N) = 0.$$

Hence we have (5.4).

Given $\epsilon > 0$, by (5.4) we have

$$\bigcap_{N=1}^{\infty} \left\{ \{x \in B; ||x|| \geq \epsilon\} \cap Q_N(K) \right\} = \phi.$$

Note that K is compact in B from (i) and Lemma 3.1, $Q_N(K) \subset K$ for all $N \geq 1$ and Q_N is a closed continuous operator for each $N \geq 1$. Consequently, for each $N \geq 1$ the set $\{x \in B; ||x|| \geq \epsilon\} \cap Q_N(K)$ is compact in B. By (5.3) and the finite intersection property of compact sets, there exists $N_o \geq 1$ such that

$$\{x \in B; ||x|| \geq \epsilon\} \cap Q_{N_o}(K) = \phi.$$

Equivalently,

$$Q_{N_o}(K) \subset \{x \in B; ||x|| < \epsilon\}.$$

Hence we have (5.2). ∎

Theorem 4.3 takes some simpler forms when the chain $\{X_n\}_{n \geq 0}$, the map ξ or the value space B satisfies certain additional conditions. In the discussion below we give some simplified versions of Theorem 4.3 under certain specifications. First, obverve that $\xi \in WCLT$ implies the condition (iii') in Theorem 4.3 when B is a finite dimensional space. Therefore, the problem raised at the end of Section III-2 is answered in a fairly satisfactory way. More generally, this simplification, like in the independent context, can be extended to the case when B is a type 2 space. Recall (Chapter 9, Ledoux-Talagrand (1991)) that a separable Banach space B is said to be of type p $(1 \leq p \leq 2)$ if there is a constant M such that for all finite sequences $\{x_i\}$ in B,

$$E||\sum_i \epsilon_i x_i||^p \leq M \sum_i ||x_i||^p,$$

where $\{\epsilon_i\}$ is a Rademacher sequence. By definition, one can easily see that all separable Hilbert spaces are type 2 spaces.

Theorem 5.2. *Let $\{X_n\}_{n\geq 0}$ be an ergodic Markov chain and let B be a type 2 space. Assume (4.3) holds. Then, the bounded LIL holds if and only if the following two conditions are fulfilled:*

(i) $$\xi \in WCLT,$$

(ii) for some (all) small set C,

$$\int_C \pi(dx) E_x \max_{n \leq \tau_C} \left(||S_n||^2 / L_2 ||S_n|| \right) < +\infty,$$

in which case we have

$$\limsup_{n \to \infty} \frac{||S_n||}{\sqrt{2nL_2 n}} = \sigma \quad a.s.$$

$$\lim_{n \to \infty} d(S_n/\sqrt{2nL_2 n}, K) = 0 \quad a.s.$$

and

$$C\left(\{S_n/\sqrt{2nL_2 n}\}_{n\geq 1}\right) = K \quad a.s.$$

Proof. Our assertion follows in the proof of Theorem 4.3, except the fact that (4.36) follows from (4.30) and Lemma 8.7 in Ledoux-Talagrand (1991). ∎

Another possibility is to simplify the conditions (i) and (ii) in Theorem 4.3 by imposing some assumptions on the chain $\{X_n\}_{n\geq 0}$ and the map ξ. Recall (Theorem II-4.1) that under the ergodicity of degree 2, the CLT holds for all centered bounded function from E to $\mathbf{R}$. The following result shows this is also true for the LIL.

Theorem 5.3. *Let $\{X_n\}_{n\geq 0}$ be ergodic of degree 2 and let the map $\xi: E \longrightarrow B$ be bounded and satisfy*

$$\int f \circ \xi(x)\pi(dx) = 0 \quad \forall f \in B^*.$$

Then, $\xi \in WCLT$ with the g.c.f. $\{\sigma_f^2;\ f \in B^*\}$ given by

$$\sigma_f^2 = \int (f \circ \xi)^2(x)\pi(dx) + 2\int \sum_{n=1}^{\infty} f \circ \xi(x) P^n(f \circ \xi)(x)\pi(dx) \quad f \in B^*.$$

Morever,

$$\limsup_{n\to\infty} \frac{\|S_n\|}{\sqrt{2nL_2n}} < +\infty \quad a.s.,$$

provided

$$\{S_n/\sqrt{2nL_2n}\}_{n\geq 1} \quad \text{is bounded in probability.}$$

Further,

$$\limsup_{n\to\infty} \frac{\|S_n\|}{\sqrt{2nL_2n}} = \sigma \quad a.s.,$$

$$\lim_{n\to\infty} d(S_n/\sqrt{2nL_2n}, K) = 0 \quad a.s.$$

and

$$C\Big(\{S_n/\sqrt{2nL_2n}\}_{n\geq 1}\Big) = K \quad a.s.,$$

provided

$$S_n/\sqrt{2nL_2n} \longrightarrow 0 \quad \text{in probability.}$$

Proof. The WCLT part comes from Theorem II-4.1 (with $f \circ \xi$ instead of ξ for each $f \in B^*$). The rest follows from Theorem 4.3 with the following observation:

$$\int_C \pi(dx) E_x \max_{n \leq \tau_C} \left(||S_n||^2 / L_2 ||S_n|| \right) \leq \int_C \pi(dx) E_x \max_{n \leq \tau_C} \left(||S_n||^2 \right)$$

$$\leq ||\xi||_\infty^2 \int_C \pi(dx) E_x \tau_C^2 = ||\xi||_\infty^2 \left(2 \int \pi(dx) E_x \tau_C - 1 \right) < +\infty,$$

where the equality comes from Proposition 5.15 in Nummelin (1984). ∎

Comparing the next result with the Ledoux-Talagrand LIL (see Chapter 8 in [31] for details), we will see that as far as the LIL is concerned, functionals of uniformly ergodic Markov chain behave like i.i.d. sequences.

Theorem 5.4. *Let $\{X_n\}_{n \geq 0}$ be an uniformly ergodic Markov chain and assume*

$$\int f \circ \xi(x) \pi(dx) = 0 \quad \text{and} \quad \int (f \circ \xi)^2 (x) \pi(dx) < +\infty \quad \forall f \in B^*.$$

Then, $\xi \in WCLT$ with the g.c.f. $\{\sigma_f^2; \ f \in B^*\}$ given by

$$\sigma_f^2 = \int (f \circ \xi)^2(x) \pi(dx) + 2 \int \sum_{n=1}^\infty f \circ \xi(x) P^n (f \circ \xi)(x) \pi(dx) \quad f \in B^*.$$

Morever, in order that

$$\limsup_{n \to \infty} \frac{||S_n||}{\sqrt{2n L_2 n}} < +\infty \quad a.s.,$$

it is sufficient and necessary that the following two conditions are fulfilled:

(5.5) $$\int \left(||\xi(x)||^2 / L_2 ||\xi(x)|| \right) \pi(dx) < +\infty$$

and

$$\left\{ S_n / \sqrt{2n L_2 n} \right\}_{n \geq 1} \quad \text{is bounded in probability.}$$

Further,

$$\limsup_{n\to\infty} \frac{\|S_n\|}{\sqrt{2nL_2n}} = \sigma \quad a.s.,$$

$$\lim_{n\to\infty} d(S_n/\sqrt{2nL_2n}, K) = 0 \quad a.s.$$

and

$$C\Big(\{S_n/\sqrt{2nL_2n}\}_{n\geq 1}\Big) = K \quad a.s.,$$

provided (5.5) holds and

$$S_n/\sqrt{2nL_2n} \longrightarrow 0 \quad \text{in probability.}$$

Proof. The WCLT part comes from Theorem II-4.3 (with $f \circ \xi$ instead of ξ for each $f \in B^*$). The rest follows from Theorem 4.3 and the fact that when $\{X_n\}_{n\geq 0}$ is uniformly ergodic, the state space E is a small set (Theorem 16.02 in Meyn-Tweedie (1993)). Thus (5.5) is equivalent to the condition (ii) in Theorem 4.3. ∎

III-6. Supplement

The method used in establishing the LIL in this chapter offers a new way to look at other strong limit problems in Markovian context. We anticipate that more applications will be found in future study. In this section we make some remarks on the argument used in this chapter.

In view of the proof of Theorem 2.2, the reader might wonder if the resolvent approximation argument is applicable in the vector setting. If so, the proof of Theorem 4.3 could be reduced to the situation in which $\{X_n\}_{n\geq 0}$ has an atom. The following example shows that the resolvent approximation is no longer effective in the Banach valued case.

Example 6.1. Let c_o be the separable Banach space of all real sequences tending to zero endowed with the norm $||\cdot||$ given by

$$||x|| = \sup_{k\geq 1} |x_k|, \quad \text{where } x = \{x_k\}_{k\geq 1} \in c_o.$$

Let $\{Z_n\}_{n\geq 0}$ be an i.i.d. c_o-valued random sequence with Z_0 defined by

$$Z_0 = \left\{\frac{\epsilon_k}{\sqrt[4]{\log(k+1)}}\right\}_{k\geq 1},$$

where $\{\epsilon_k\}_{k\geq 1}$ is a Rademacher sequence. Clearly, Z_0 is bounded and symmetric. By an example in Chen ([14]),

$$\limsup_{n\to\infty} \Big|\Big|\sum_{k=0}^{n-1} Z_k\Big|\Big|/\sqrt{2nL_2n} = +\infty \quad \text{a.s.}$$

Comparing this with the Ledoux-Talagrand LIL (Theorem 8.6, [31]), we must have

$$\Big\{\sum_{k=0}^{n-1} Z_k/\sqrt{2nL_2n}\Big\}_{n\geq 1} \quad \text{is unbounded in probability.}$$

Consequently,

(6.1) $$\limsup_{n\to\infty} E\Big|\Big|\sum_{k=0}^{n-1} Z_k\Big|\Big|/\sqrt{2nL_2n} = +\infty.$$

Let $E = c_o \times c_o$ and define

$$X_n = (Z_n, Z_{n+1}) \quad n = 0, 1, 2, \cdots.$$

One can easily see that $\{X_n\}_{n\geq 0}$ is a Markov chain with state space E, transition P given by

$$P((x,y),\cdot) = (\delta_y \otimes \mu)(\cdot) \quad (x,y) \in c_o \times c_o$$

and the initial distribution $\pi = \mu \otimes \mu$, where $\mu = \mathcal{L}(Z_0)$. Morever, $\{X_n\}_{n\geq 0}$ is (uniformly) ergodic (Example I-1.1). Let $\xi: c_o \times c_o \longrightarrow c_o$ be defined by $\xi(x,y) = x - y$, where $(x,y) \in c_o \times c_o$. Then,

$$\|\xi(X_0)\| \leq \frac{2}{\sqrt[4]{log 2}} \quad \text{a.s.}$$

and

$$S_n = \sum_{k=0}^{n-1} \xi(X_k) = Z_n - Z_0 \quad (n \geq 1).$$

One can easily and directly verify that all conditions and all conclusions in Theorem 4.3 are satisfied with the g.c.f. $\sigma_f^2 \equiv 0$ and $C = E$. On the other hand we claim that

(1). The chain $\{X_n\}_{n\geq 0}$ has no small set of order 1.

(2). For each $0 < t < 1$, let $\{X_n^t\}_{n\geq 0}$ be a resolvent chain of $\{X_n\}_{n\geq 0}$ with the transition P_t given in (II-3.6). Then,

(6.2) $$\limsup_{n\to\infty} \|\sum_{k=0}^{n-1} \xi(X_n^t)\|/\sqrt{2nL_2n} = +\infty \quad \text{a.s.}$$

By Example I-1.1, the first assertion will hold if

(6.3) $$\mu(\{x\}) = 0 \quad \forall x \in c_o.$$

By definition, we need only to verify (6.3) for the point $x' \in c_o$ of the form:

$$x' = \left\{\frac{\pm 1}{\sqrt[4]{log(k+1)}}\right\}_{k\geq 1}.$$

For each $n \geq 1$, let

$$A_n = \{x \in c_o;\ x \text{ coincides with } x' \text{ in the first } n \text{ coordinates}\}.$$

Then,
$$\mu(\{x'\}) \le \mu(A_n) = 2^{-n} \longrightarrow 0 \quad (n \to \infty).$$

We now prove the assertion (2). Fix $0 < t < 1$ and let the random sequences $\{\delta_k\}_{k \ge 1}$ and $\{T_k\}_{k \ge 0}$ be given in (II-3.3) and (II-3.4), respectively. Then,

(6.4)
$$\sum_{k=1}^{n} \delta_k \xi(X_k) = \sum_{k=1}^{n} \delta_k (Z_{k+1} - Z_k)$$
$$= \delta_n Z_{n+1} - \delta_1 Z_1 + \sum_{k=2}^{n} (\delta_{k-1} - \delta_k) Z_k.$$

By the symmetry of $\{Z_n\}$,
$$E\|\sum_{k=2}^{n} (\delta_{k-1} - \delta_k) Z_k\| = E\|\sum_{k=2}^{n} |\delta_{k-1} - \delta_k| Z_k\|$$
$$\ge E|\delta_1 - \delta_2| \cdot E\|\sum_{k=2}^{n} Z_k\| = 2t(1-t) E\|\sum_{k=2}^{n} Z_k\| \quad (n \ge 2),$$

where the inequality follows from the Jessen's inequality. By (6.1), therefore,
$$\limsup_{n \to \infty} E\|\sum_{k=2}^{n} (\delta_{k-1} - \delta_k) Z_k\|/\sqrt{2nL_2 n} = +\infty.$$

Applying Proposition 6.8 of [31] conditionally on $\{\delta_k\}_{k \ge 1}$ gives that
$$\{\sum_{k=2}^{n} (\delta_{k-1} - \delta_k) Z_k / \sqrt{2nL_2 n}\}_{n \ge 1} \quad \text{is unbounded in probability.}$$

In view of (6.4) we have that
$$\{\sum_{k=1}^{n} \delta_k \xi(X_k) / \sqrt{2nL_2 n}\}_{n \ge 1} \quad \text{is unbounded in probability.}$$

Consequently,
$$\limsup_{n \to \infty} \|\sum_{k=1}^{n} \delta_k \xi(X_k)\|/\sqrt{2nL_2 n} = +\infty \quad \text{a.s.}$$

From the definitions of $\{\delta_k\}_{k\geq 1}$ and $\{T_k\}_{k\geq 0}$,

$$\max_{k\leq n}\Big\|\sum_{j=1}^{T_k}\delta_k\xi(X_j)\Big\| = \max_{k<T_n}\Big\|\sum_{j=1}^{k}\delta_k\xi(X_j)\Big\| \geq \Big\|\sum_{k=1}^{n}\delta_k\xi(X_k)\Big\|.$$

Hence

$$\limsup_{n\to\infty}\Big\|\sum_{k=1}^{T_n}\delta_k\xi(X_k)\Big\|\Big/\sqrt{2nL_2 n} = +\infty \quad \text{a.s.}$$

By Lemma II-3.3, this is equivalent to (6.2). ∎

Using a similar argument, we can obtain other limit theorems for the ergodic Markov chains. As an example, we obtain a theorem on the law of the large numbers, which extends Theorem 3.1 in de Acosta (1981) from the i.i.d. case to the Markovian case. We give it in the following without proof, which is similar to that of Theorem 4.3.

Theorem 6.2. *Let $\{X_n\}_{n\geq 0}$ be an ergodic Markov chain and let $1 \leq p < 2$. Then the following two statements (i) and (ii) are equivalent:*

(i) For some (all) small set C,

(6.5) $$\int_C \pi(dx) E_x \max_{n\leq \tau_C}\|S_n\|^p < +\infty$$

and

(6.6) $$S_n/n^{1/p} \longrightarrow 0 \quad \text{in probability.}$$

(ii) The Marcinkiewicz-Zygmund's law of the large numbers holds, i.e.,

$$\lim_{n\to\infty} S_n/n^{1/p} = 0 \quad \text{a.s.}$$

Similarly, one can see that (6.5) and the condition

$$\int \xi(x)\pi(dx) = 0,$$

where the integration is defined in Pettis sense, imply (6.6) when B is a type p space. Hence Theorem 6.2 takes simpler form in this case. In particular, Theorem 6.2 generalizes the law of the large numbers of of Marcinkiewicz-Zygmund if we take $B = \mathbf{R}$. Comparing Theorem 6.2 with Example 2.1 in this case we conclude that in that example, (6.5) must be violated. In fact, one can easily and directly verify that in Example 2.1,

$$E_0 \max_{n \leq \tau_0} |S_n|^p = +\infty.$$

Chapter IV. The Moderate Deviation Principle

IV-1. Introduction

In this chapter, B is a separable Banach space with norm $\|\cdot\|$ and topological dual B^* whose closed unit ball is denoted by B_1^*. Let $\{X_n\}_{n\geq 0}$ be an ergodic Markov chain and let $\xi\colon E \longrightarrow B$ be a measurable map. Set

$$S_n = \sum_{k=0}^{n-1} \xi(X_k) \qquad n=1,2,\cdots.$$

Throughout, $\{b_n\}_{n\geq 1}$ is a non-decreasing sequence of positive numbers such that

(1.1) $\qquad b_n/\sqrt{n} \longrightarrow \infty \quad \text{and} \quad b_n/n \longrightarrow 0 \quad (n\to\infty).$

The goal of this chapter is to study the asymptotics of the probabilities

$$P_\mu\{S_n/b_n \in A\} \qquad n=1,2,\cdots$$

as $n\to\infty$, where A is a Borel subset of B. More precisely, we shall prove under certain conditions that

(1.2) $\qquad \limsup_{n\to\infty} \dfrac{n}{b_n^2} \log P_\mu\!\left\{\dfrac{S_n}{b_n} \in F\right\} \leq -\inf_{x\in F} I(x)$

for all closed subset F of B, and

(1.3) $\qquad \liminf_{n\to\infty} \dfrac{n}{b_n^2} \log P_\mu\!\left\{\dfrac{S_n}{b_n} \in G\right\} \geq -\inf_{x\in G} I(x)$

for all open subset G of B, where the rate function $I\colon B \longrightarrow [0,+\infty]$ will be specified in the sequel. This type of limit is called a moderate deviation.

The previous literature on moderate deviations for Markov chains can be found in [4], [5], [24], [34], [44], where the focus is on either the empirical measures or finite dimensional functionals. In [24], [34], [44], the upper and lower bounds are established under the same conditions. This makes the results not so optimal since generally, the upper bound needs substantially more restrictive assumptions than the lower bound does. Noticing such a difference, de Acosta ([4]) obtains the lower bounds for empirical measures and bounded, centered, finite dimensional functionals of Markov chains under the assumption of ergodicity of degree 2. Since the definition of the rate function in the empirical measure context needs that the CLT hold for all bounded, centered and real valued functions on E, the sharpness of such a result is obvious in view of Theorem II-4.1. In [5], upper bounds for empirical measures and bounded, centered and finite dimensional functionals of Markov chains are obtained under the assumption of geometric ergodicity. This assumption is in a certain sense shown to be best possible for the upper bounds (see Remark 5.1 in [5]).

Our situation is quite different from those in the previous literature. On the one hand, the infinite dimensionality of B gives rise to new difficulties. On the other hand, rather than considering all bounded, centered and finite dimensional functionals on E, we study the moderate deviation for single (sometimes unbounded) map ξ, in which case the ergodicity of degree 2 is neither necessary nor sufficient for the definition of the rate function. An interesting question in this case is how to establish the lower bound under the "basic" ergodicity assumption together with some other conditions imposed on the map ξ.

To define the rate function I which governs the moderate deviation we address, we assume that $\xi \in WCLT$ with the Gaussian covariance function (g.c.f.) $\{\sigma_f^2;\ f \in B^*\}$. Then I is defined by

(1.4) $$I(x) = \sup_{f \in B^*} \{f(x) - \frac{1}{2}\sigma_f^2\} \quad x \in B.$$

One can see that $I \geq 0$ and I is lower semi-continuous on B. Hence for each $l \geq 0$, the set

(1.5) $$\{x;\ I(x) \leq l\}$$

is closed in B. Recall that the g.c.f. $\{\sigma_f^2;\ f \in B^*\}$ can be represented in an explicit way under some mild conditions therefore so can I through (1.4). Indeed, by Theorem II-3.1 (with $f \circ \xi$ instead of ξ for each $f \in B^*$), we have $\xi \in WCLT$ with the g.c.f. $\{\sigma_f^2;\ f \in B^*\}$ given by

(1.6) $$\sigma_f^2 = \int (f \circ \xi)^2(x)\pi(dx) + 2\int \sum_{n=1}^{\infty} f \circ \xi(x) P^n(f \circ \xi)(x) \pi(dx) \quad f \in B^*$$

provided

(1.7) $$\int (f \circ \xi)^2(x)\pi(dx) < +\infty \quad f \in B^*$$

and

(1.8) $$\sum_{n=1}^{\infty} f \circ \xi(\cdot) P^n(f \circ \xi)(\cdot) \quad \text{converges in } L^1(\pi) \text{ for each } f \in B^*.$$

Another way to represent the rate function I is to write it in terms of the reproducing kernel Hilbert space $(H, \|\cdot\|_H)$ generated by $\{\sigma_f^2;\ f \in B^*\}$, which is constructed in Section III-3 and always exists under the conditions given in all theorems in this chapter. In view of (III-3.3), we have (Lemma 3.44, Stroock (1984)) that

(1.9) $$I(x) = \begin{cases} \frac{1}{2}\|x\|_H^2 & x \in H \\ +\infty & x \notin H. \end{cases}$$

According to Lemma III-3.1-(iii), the rate function I is good, by which we mean that the set given in (1.5) is compact in B for each $l \geq 0$, if and only if $\{\sigma_f^2;\ f \in B^*\}$ is weak-star sequentially continuous.

We study the upper and lower bounds separately in this chapter. Here is our plan: The upper bounds are given in Section IV-2. The assumptions for the upper bounds appear to be best possible for the result and they are given in a very explicit way so that one can see how the tail behaviors of S_n depend on the degree of ergodicity of the Markov chain. Based on the "basic" ergodicity assumption, three obviously incomparable theorems on the lower bounds are established in Section IV-3, IV-4 and IV-5, respectively. The proofs mainly depend on the split and regeneration scheme developed in Chapter I and the results on the central limit theorem obtained in Section II-2. However, we would like to mention that some ideas in the arguments for the lower bounds come from elsewhere despite that the main concerns are different: The technique of Lévy decomposition we will use was elegantly applied, for the first time, to the strong limit theorems by de Acosta and Kuelbs (1983). This nice idea was adopted later by the author ([11], [14]) in the study of the lower bounds of moderate deviations in the independent context and related problems. The method of directly using the minorization of the Markov kernel, which is effectively exploited in Dinwoodie-Ney ([22]) and de Acosta ([4]), and the idea of dividing the sum S_n in de Acosta ([4]) are adopted and further developed here in such a way that we divide S_n at the exact moment the chain visits a fixed small set rather than any deterministic time. The difference made by this improvement is that the map ξ no longer need to be bounded as soon as the central limit theorem holds. An isoperimetric inequality due to Ledoux and Talagrand (1991) is also introduced. The paper [14] provides some useful ideas on how to apply this powerful inequality to the lower bound problems.

The incomparability of the three theorems on the lower bounds makes it desirable to know if they can be unified into one result. Specifically, we conjecture that the minorization condition (3.1) in Theorem 3.1 might be removable. We leave this to a future study.

IV-2. Upper bounds

One of the assumptions for the upper bounds is geometric ergodicity. Recall that the Markov chain $\{X_n\}_{n\geq 0}$ is called geometrically ergodic if it is ergodic and there exists $r_o > 1$ such that

$$(2.1) \qquad \sum_{n=1}^{\infty} r_o^n \int ||P^n(x,\cdot) - \pi||_V \pi(dx) < +\infty.$$

Since geometric ergodicity implies ergodicity of degree 2, by Theorem II-4.1 we have that $\xi \in WCLT$ for any measurable and bounded map $\xi \colon E \longrightarrow B$ with the g.c.f. $\{\sigma_f^2;\ f \in B^*\}$ defined by (1.6). Applying the dominated convergence theorem to (1.6) one can easily see how (2.1) implies the weak-star sequential continuity of $\{\sigma_f^2;\ f \in B^*\}$. Therefore, we have the goodness of the rate function given in (1.4).

Here is the main result in this section:

Theorem 2.1. *Let $\{X_n\}_{n\geq 0}$ be a geometrically ergodic Markov chain and let $\xi \colon E \longrightarrow B$ be a measurable and bounded map. Let $\{b_n\}_{n\geq 1}$ be a non-decreasing sequence of positive numbers satisfying (1.1) and assume that*

$$(2.2) \qquad S_n/b_n \longrightarrow 0 \quad \text{in probability}.$$

Then, for any initial distribution μ satisfying

$$(2.3) \qquad \sum_{n=1}^{\infty} r_\mu^n ||\mu P^n - \pi||_V < +\infty \quad \text{for some}\ \ r_\mu > 1$$

and for any closed subset F of B,

(2.4) $$\limsup_{n\to\infty} \frac{n}{b_n^2} \log P_\mu\left\{\frac{S_n}{b_n} \in F\right\} \leq -\inf_{x\in F} I(x).$$

In particular, (2.4) holds for $\mu = \pi$.

Let us make some comments before the proof. The assumption of geometric ergodicity and the condition (2.3) are best possible for the upper bound (2.4) in a sense given in de Acosta-Chen (Remark 5.1, [5]). The condition (2.2) is standard in the infinite dimensional context. From (III-3.6) one has that
$$\inf_{||x||\geq\epsilon} I(x) > 0 \quad \forall \epsilon > 0.$$
Therefore, the condition (2.2) is necessary for the upper bound (2.4).

To prove Theorem 2.1, we need the following two lemmas. The first one shows how the condition (2.3) is used in our proof.

Lemma 2.2. *Let $\{X_n\}_{n\geq 0}$ be a geometrically ergodic Markov chain. Then, for any probability measure μ on $(E, \mathcal{E})$, the condition (2.3) implies the geometrical regularity of μ: Given $A \in \mathcal{E}^+$, there exists $\lambda > 1$ depending only on μ and A such that*

(2.5) $$E_\mu \lambda^{\tau_A} < +\infty.$$

Proof. Let $A \in \mathcal{E}^+$ be fixed. By Proposition 5.19-(i) in Nummelin (1984), there is $\lambda_o > 1$ such that

(2.6) $$E_\pi \lambda_o^{\tau_A} < +\infty.$$

For $n \geq 1$ and $x \in E$,
$$P_\mu\{\tau_A \geq 2n\} \leq P_\mu\{X_{n+1} \notin \alpha, \cdots, X_{2n-1} \notin \alpha\}$$
$$= \int \mu P^n(dx) P_x\{\tau_A \geq n\} \leq ||\mu P^n - \pi||_V + P_\pi\{\tau_A \geq n\}.$$

Let $\lambda = \min\{\sqrt{r_\mu}, \sqrt{\lambda_o}\}$. It follows from (2.3) and (2.6) that

$$\sum_{n=1}^{\infty} \lambda^{2n} P_\mu\{\tau_A \geq 2n\} < +\infty,$$

which implies (2.5). ∎

The next lemma is an extension of Ottavanii's inequality to the 1-dependent random variables which will be used to approximate the Markov chain by the regeneration argument.

Lemma 2.3. *Let $\xi_1, \xi_2, \cdots, \xi_n$ ($n \geq 3$) be B-valued random variables such that for every $1 < k < n$, the following two collections*

$$\{\xi_1, \cdots, \xi_{k-1}\} \text{ and } \{\xi_{k+1}, \xi_{k+2}, \cdots, \xi_n\}$$

are independent. Then, for any $r, s, t > 0$,

(2.7)
$$P\Big\{\max_{k \leq n} \Big\|\sum_{j=1}^{k} \xi_j\Big\| \geq r + s + t\Big\}$$
$$\leq (1-c)^{-1}\Big(P\Big\{\Big\|\sum_{j=1}^{n} \xi_j\Big\| \geq t\Big\} + P\Big\{\max_{k \leq n} \|\xi_j\| \geq s\Big\}\Big),$$

where we assume

$$c \equiv \max_{k \leq n-2} P\Big\{\Big\|\sum_{j=k+2}^{n} \xi_j\Big\| \geq r\Big\} < 1.$$

The proof is standard. For the sake of completeness we include it here.

Proof. Let

$$T = \inf\Big\{k \geq 1; \ \Big\|\sum_{j=1}^{k} \xi_j\Big\| \geq r + s + t\Big\}.$$

Then,

$$P\left\{\left\|\sum_{j=1}^{n}\xi_j\right\|\geq t\right\}$$

$$\geq P\left\{T\leq n,\ \left\|\sum_{j=1}^{n}\xi_j\right\|\geq t\right\}$$

$$=\sum_{k=1}^{n}P\left\{T=k,\ \left\|\sum_{j=1}^{n}\xi_j\right\|\geq t\right\}$$

$$\geq \sum_{k=1}^{n}P\left\{T=k,\ \left\|\sum_{j=k+2}^{n}\xi_j\right\|\leq r\ \|\xi_{k+1}\|\leq s\right\}$$

$$\geq \sum_{k=1}^{n}P\{T=k\}\cdot P\left\{\left\|\sum_{j=k+2}^{n}\xi_j\right\|\leq r\right\}-\sum_{k=1}^{n}P\{T=k,\ \|\xi_{k+1}\|>s\}$$

$$\geq (1-c)P\{T\leq n\}+P\left\{\max_{k\leq n}\|\xi_j\|\geq s\right\}$$

$$=(1-c)P\left\{\max_{k\leq n}\left\|\sum_{j=1}^{k}\xi_j\right\|\geq r+s+t\right\}+P\left\{\max_{k\leq n}\|\xi_j\|\geq s\right\}.$$

This is (2.7). ∎

Proof of Theorem 2.1. Let the small set C be fixed. By definition there exists $m\geq 1$, $b>0$ and a probability measure ν on $(E,\mathcal{E})$ such that

(2.8) $$P^m\geq bI_C\otimes \nu.$$

Consider the split chain $\tilde{P}$ and let the regeneration times $\{\tau(k)\}_{k\geq 0}$ be defined as in (I-2.19) and $i(n)$ as in (I-3.10). Recall that for each $n\geq 1$

(2.9) $$S_n = S_{m(\tau(0)+1)\wedge n}+\sum_{k=1}^{i(n)-1}\xi_k+\sum_{j=m(l(n)+1)}^{n-1}\xi(X_j)\quad \text{a.s.,}$$

where the 1-dependent B-valued random sequence $\{\xi_k\}_{k\geq 1}$ is given by

$$\xi_k = \sum_{j=m(\tau(k-1)+1)}^{m\tau(k)+m-1}\xi(X_j)\quad k=1,2,\cdots$$

and

$$l(n) = \tau((i(n) - 1) \vee 0)$$
$$= \begin{cases} \max\{k \geq 0;\ k \leq \left[\frac{n}{m}\right] - 1 \text{ and } \Phi_n \in \alpha^*\} & \text{if } \tau(0) \leq \left[\frac{n}{m}\right] - 1 \\ \tau(0) & \text{if } \tau(0) > \left[\frac{n}{m}\right] - 1. \end{cases}$$

By Theorem I-2.5, $\{\Phi_n\}_{n\geq 0}$ is geometrically ergodic and (2.3) is equivalent to

$$\sum_{n=1}^{\infty} r_\mu^n \|\mu^* \bar{P}^n - \pi^*\|_V < +\infty.$$

Applying Lemma 2.2 to the chain $\{\Phi_n\}_{n\geq 0}$ gives

(2.10) $$\tilde{E}_{\mu^*} \lambda_\mu^{\tau(0)} < +\infty$$

for some $\lambda_\mu > 1$. In particular, since from (2.8),

$$\pi(\cdot) = \int \pi(dx) P^m(x, \cdot) \geq \int_C \pi(dx) P^m(x, \cdot) \geq b\pi(C) \cdot \nu(\cdot),$$

and consequently,

$$\|\nu P^n - \pi\|_V \leq \int \nu(dx) \|P^n(x, \cdot) - \pi\|_V$$
$$\leq b^{-1} \pi(C)^{-1} \int \pi(dx) \|P^n(x, \cdot) - \pi\|_V \quad (n \geq 1).$$

Hence by geometric ergodicity, ν satisfies the condition (2.3). Therefore,

(2.11) $$\tilde{E}_{\nu^*} \lambda_\nu^{\tau(0)} < +\infty$$

for some $\lambda_\nu > 1$. By assumption, $\|\xi\|_\infty < +\infty$. Hence (2.11) implies (Corollary I-2.4) that

(2.12) $$E \exp\{\epsilon_o \|\xi_1\|\} < +\infty$$

for some $\epsilon_o > 0$. In view of (2.2), one must have $\int \xi(x)\pi(dx) = 0$. By Theorem 10.4.9 in Meyn-Tweedie (1993),

(2.13) $$E\xi_1 = 0.$$

To control each term on the right-hand side of (2.9), we now prove that for any $\epsilon > 0$,

(2.14) $$\lim_{n\to\infty} \frac{n}{b_n^2} \log \tilde{P}_{\mu^*}\{||S_{m(\tau(0)+1)\wedge n}|| \geq \epsilon b_n\} = -\infty,$$

(2.15) $$\lim_{n\to\infty} \frac{n}{b_n^2} \log \tilde{P}_{\mu^*}\left\{\Big\|\sum_{j=m(l(n)+1)}^{n-1} \xi(X_j)\Big\| \geq \epsilon b_n\right\} = -\infty,$$

(2.16) $$\lim_{n\to\infty} \frac{n}{b_n^2} \log \tilde{P}_{\mu^*}\left\{\Big\|\sum_{k=1}^{i(n)-1} \xi_k - \sum_{k=1}^{e(n)} \xi_k\Big\| \geq \epsilon b_n\right\} = -\infty,$$

where
$$e(n) = [\pi^*(\alpha^*)nm^{-1}] \quad n = 1, 2, \cdots.$$

In fact, since
$$||S_{m(\tau(0)+1)\wedge n}|| \leq \sum_{j=0}^{m\tau(0)+m-1} ||\xi(X_j)|| \leq (m\tau(0) + m - 1)||\xi||_\infty,$$

the claim (2.14) follows from (2.10). For (2.15), we use the following estimation:
$$\tilde{P}_{\mu^*}\left\{\Big\|\sum_{j=m(l(n)+1)}^{n-1} \xi(X_j)\Big\| \geq \epsilon b_n\right\}$$
$$\leq \tilde{P}_{\mu^*}\left\{(n-1-m(l(n)+1))||\xi||_\infty \geq \epsilon b_n\right\}$$
$$\leq \tilde{P}_{\mu^*}\left\{l(n) \leq (m/n) - m^{-1}||\xi||_\infty^{-1}\epsilon b_n - 1\right\}$$
$$\leq \sum_{k=[m^{-1}||\xi||_\infty^{-1}\epsilon b_n]}^{[n/m]} \tilde{P}_{\mu^*}\{l(n) = [n/m] - k\}$$

LIMIT THEOREMS FOR FUNCTIONALS

$$= \sum_{k=[m^{-1}\|\xi\|_\infty^{-1}\epsilon b_n]}^{[n/m]} \tilde{P}_{\mu^*}\{\Phi_{[n/m]-k} \in \alpha^*, \ \Phi_{[n/m]-k+1} \notin \alpha^*,$$

$$\ldots, \Phi_{[n/m]-1} \notin \alpha^*\}$$

$$= \sum_{k=[m^{-1}\|\xi\|_\infty^{-1}\epsilon b_n]}^{[n/m]} \tilde{P}_{\mu^*}\{\Phi_{[n/m]-k} \in \alpha^*\}\tilde{P}_{\alpha^*}\{\tau_{\alpha^*} \geq k\}$$

$$\leq \sum_{k=[m^{-1}\|\xi\|_\infty^{-1}\epsilon b_n]}^{\infty} \tilde{P}_{\nu^*}\{\tau(0) \geq k-1\}$$

Therefore, (2.15) follows from (2.11).

We now prove (2.16). Let $0 < \delta < m^{-1}\pi^*(\alpha^*)$ be fixed but arbitrary, and let n sufficiently large such that $e(n) \geq \delta n$. Define

$$k(n) = \inf\{k; \ k \geq e(n) - \delta n\}.$$

Notice that on the event $\{|i(n) - 1 - e(n)| \leq \delta n\}$, we have

$$\left\|\sum_{k=1}^{i(n)-1}\xi_k - \sum_{k=1}^{e(n)}\xi_k\right\| \leq 2\max_{|k-1-e(n)|\leq \delta n}\left\|\sum_{j=k(n)}^{k}\xi_j\right\|.$$

By the stationarity of $\{\xi_k\}_{k\geq 1}$ we have

(2.17)
$$\tilde{P}_{\mu^*}\left\{\left\|\sum_{k=1}^{i(n)-1}\xi_k - \sum_{k=1}^{e(n)}\xi_k\right\| \geq \epsilon b_n\right\}$$
$$\leq \tilde{P}_{\mu^*}\{|i(n) - 1 - e(n)| > \delta n\}$$
$$+ P\left\{\max_{k\leq 2\delta n}\left\|\sum_{j=1}^{k}\xi_j\right\| \geq \frac{\epsilon}{2}b_n\right\}$$
$$\leq \tilde{P}_{\mu^*}\{i(n) \geq e(n) + [\delta n] + 1\}$$
$$+ \tilde{P}_{\mu^*}\{i(n) < e(n) - [\delta n] + 1\}$$
$$+ P\left\{\max_{k\leq 2\delta n}\left\|\sum_{j=1}^{k}\xi_j\right\| \geq \frac{\epsilon}{2}b_n\right\}.$$

By definition,

$$\begin{aligned}
&\tilde{P}_{\mu^*}\{i(n) \geq e(n) + [\delta n] + 1\} \\
&\leq \tilde{P}_{\mu^*}\{\tau(e(n) + [\delta n]) \leq \left[\frac{n}{m}\right] - 1\} \\
&\leq \tilde{P}_{\mu^*}\{\tau(e(n) + [\delta n]) - \tau(0) \leq \left[\frac{n}{m}\right] - 1\} \\
&= \tilde{P}_{\mu^*}\left\{\sum_{k=1}^{e(n)+[\delta n]} (\tau(k) - \tau(k-1)) \leq \left[\frac{n}{m}\right] - 1\right\} \\
&= \tilde{P}_{\mu^*}\left\{\frac{1}{e(n) + [\delta n]} \sum_{k=1}^{e(n)+[\delta n]} (\tau(k) - \tau(k-1)) \right. \\
&\qquad\left. \leq \left(\left[\frac{n}{m}\right] - 1\right) / (e(n) + [\delta n])\right\}.
\end{aligned}$$
(2.18)

Notice that

$$\left(\left[\frac{n}{m}\right] - 1\right) / (e(n) + [\delta n]) \sim (\pi^*(\alpha^*) + \delta m)^{-1}$$
$$< \pi^*(\alpha^*)^{-1} = \tilde{E}_{\nu^*}\tau(0) = E(\tau(1) - \tau(0)),$$

and that by Theorem I-2.2, $\{\tau(k) - \tau(k-1)\}_{k \geq 1}$ is an i.i.d. sequence with the common distribution

$$\tilde{P}_{\nu^*}\{\tau(0) = j\} \quad j = 0, 1, 2, \cdots.$$

In view of (2.11), by the the upper bounds of the Cramér-Chernoff large deviations (see, e.g. Theorem 2.2.3 in [20]) and the fact (Lemma 6.1-(ii) in [6]) that the rate function $\lambda(x)$ governing the Cramér-Chernoff large deviations satisfies $\lambda(x) > 0$ for all $x \in \mathbf{R}$ but the expected value $x = \pi^*(\alpha^*)^{-1}$, there exists $u > 0$ such that

$$\tilde{P}_{\mu^*}\left\{\frac{1}{e(n) + [\delta n]} \sum_{k=1}^{e(n)+[\delta n]} (\tau(k) - \tau(k-1)) \right.$$
$$\left. \leq \left(\left[\frac{n}{m}\right] - 1\right) / (e(n) + [\delta n])\right\}$$
$$\leq e^{-un}$$

for sufficiently large n. In particular, by (2.18) one has

$$(2.19) \qquad \lim_{n \to \infty} \frac{n}{b_n^2} \log \tilde{P}_{\mu^*}\{i(n) \geq e(n) + [\delta n] + 1\} = -\infty.$$

Using a similar argument and taking (2.10) into account one can prove

$$(2.20) \qquad \lim_{n \to \infty} \frac{n}{b_n^2} \log \tilde{P}_{\mu^*}\{i(n) < e(n) - [\delta n] + 1\} = -\infty.$$

Recall ([15]) that if the 1-dependent sequence $\{\xi_k\}_{k \geq 1}$ satisfies (2.12), (2.13), and

$$(2.21) \qquad \sum_{k=1}^{n} \xi_k/b_n \longrightarrow 0 \quad \text{in probability,}$$

then for any closed subset F of B,

$$(2.22) \qquad \limsup_{n \to \infty} \frac{n}{b_n^2} P\left\{\sum_{k=1}^{n} \xi_k/b_n \in F\right\} \leq -\inf_{x \in F} J(x),$$

where

$$J(x) = \sup_{f \in B^*} \left\{f(x) - \frac{1}{2}\left(Ef^2(\xi_1) + 2Ef(\xi_1)f(\xi_2)\right)\right\}.$$

By Theorem II-2.3 (with $f \circ \xi$ instead of ξ for each $f \in B^*$),

$$(2.23) \qquad \sigma_f^2 = m^{-1}\pi^*(\alpha^*)\left(Ef^2(\xi_1) + 2Ef(\xi_1)f(\xi_2)\right).$$

From (1.4) one has

$$(2.24) \qquad J(x) = m^{-1}\pi^*(\alpha^*)I(x) \quad \forall x \in B.$$

By the argument used to establish (III-4.31) and (III-4.36) (with b_n instead of $\sqrt{2nL_2n}$) one can prove that (2.21) follows from the condition (2.2). In particular,

$$\max_{k \leq 2\delta n - 2} P\left\{\|\sum_{j=k+2}^{n} \xi_j\| \geq \frac{\epsilon}{6}b_n\right\} \leq \frac{1}{2}$$

for sufficiently large n. Applying Lemma 2.3 gives

(2.25)
$$P\Big\{ \max_{k \leq 2\delta n} \Big\| \sum_{j=1}^{k} \xi_j \Big\| \geq \frac{\epsilon}{2} b_n \Big\}$$
$$\leq 2 \Big(P\Big\{ \Big\| \sum_{j=1}^{[2\delta n]} \xi_j \Big\| \geq \frac{\epsilon}{6} b_n \Big\} + P\Big\{ \max_{k \leq 2\delta n} \|\xi_j\| \geq \frac{\epsilon}{6} b_n \Big\} \Big).$$

From (2.12) one can see that

(2.26)
$$\lim_{n \to \infty} \frac{n}{b_n^2} \log P\Big\{ \max_{k \leq 2\delta n} \|\xi_j\| \geq \frac{\epsilon}{6} b_n \Big\} = -\infty.$$

Clearly, (2.21) is equivalent to

$$\sum_{k=1}^{[2\delta n]} \xi_k / b_n \longrightarrow 0 \quad \text{in probability}.$$

Taking $F = \{x, \|x\| \geq \epsilon/6\}$ and using $[2\delta n]$ instead of n in (2.22), one has

(2.27)
$$\limsup_{n \to \infty} \frac{n}{b_n^2} \log P\Big\{ \Big\| \sum_{j=1}^{[2\delta n]} \xi_j \Big\| \geq \frac{\epsilon}{6} b_n \Big\}$$
$$\leq -(2\delta)^{-1} \inf_{\|x\| \geq \epsilon/6} J(x) = -m^{-1} \pi^*(\alpha^*)(2\delta)^{-1} \inf_{\|x\| \geq \epsilon/6} I(x),$$

where the last step follows from (2.24). From (2.25), (2.26) and (2.27),

(2.28)
$$\limsup_{n \to \infty} \frac{n}{b_n^2} \log P\Big\{ \max_{k \leq 2\delta n} \Big\| \sum_{j=1}^{k} \xi_j \Big\| \geq \frac{\epsilon}{2} b_n \Big\}$$
$$\leq -m^{-1} \pi^*(\alpha^*)(2\delta)^{-1} \inf_{\|x\| \geq \epsilon/6} I(x).$$

Combining (2.17), (2.19), (2.20) and (2.28) gives

$$\limsup_{n \to \infty} \frac{n}{b_n^2} \log \tilde{P}_{\mu^*}\Big\{ \Big\| \sum_{k=1}^{i(n)-1} \xi_k - \sum_{k=1}^{e(n)} \xi_k \Big\| \geq \epsilon b_n \Big\}$$
$$\leq -m^{-1} \pi^*(\alpha^*)(2\delta)^{-1} \inf_{\|x\| \geq \epsilon/6} I(x).$$

From (III-3.6) and (1.9) one can see that

$$\inf_{||x||>\epsilon/6} I(x) > 0.$$

Letting $\delta \longrightarrow 0^+$ gives (2.16).

We now prove (2.4). By (I-2.8) and (2.9) for given ϵ we have

(2.29)
$$\begin{aligned}
P_\mu\Big\{\frac{S_n}{b_n} \in F\Big\} &= \tilde{P}_{\mu^*}\Big\{\frac{S_n}{b_n} \in F\Big\} \\
&\leq P\Big\{\frac{1}{b_n}\sum_{k=1}^{e(n)} \xi_k \in F^{3\epsilon}\Big\} + \tilde{P}_{\mu^*}\Big\{||S_{m(\tau(0)+1)\wedge n}|| \geq \epsilon b_n\Big\} \\
&\quad + \tilde{P}_{\mu^*}\Big\{||\sum_{j=m(l(n)+1)}^{n-1} \xi(X_j)|| \geq \epsilon b_n\Big\} \\
&\quad + \tilde{P}_{\mu^*}\Big\{||\sum_{k=1}^{i(n)-1} \xi_k - \sum_{k=1}^{e(n)} \xi_k|| \geq \epsilon b_n\Big\},
\end{aligned}$$

where $F^{3\epsilon} = \{x;\ d(x, F) \leq 3\epsilon\}$. Replacing n by $e(n)$ and F by $F^{3\epsilon}$ in (2.22) gives

(2.30)
$$\begin{aligned}
\limsup_{n\to\infty} \frac{n}{b_n^2} \log P\Big\{\frac{1}{b_n}\sum_{k=1}^{e(n)} \xi_k \in F^{3\epsilon}\Big\} \\
\leq -m\pi^*(\alpha^*)^{-1} \inf_{x\in F^{3\epsilon}} J(x) = -\inf_{x\in F^{3\epsilon}} I(x),
\end{aligned}$$

where the last step follows from (2.24). Combining (2.14), (2.15), (2.16), (2.29) and (2.30) gives

$$\limsup_{n\to\infty} \frac{n}{b_n^2} \log P_\mu\Big\{\frac{S_n}{b_n} \in F\Big\} \leq -\inf_{x\in F^{3\epsilon}} I(x).$$

By Lemma 4.16 in [20] we have

$$\inf_{x\in F^{3\epsilon}} I(x) \longrightarrow \inf_{x\in F} I(x) \quad (\epsilon \to 0^+).$$

Hence (2.4) holds. ■

IV-3. Lower bounds — for chains with a small set of order 1

Let the Markov chain $\{X_n\}_{n\geq 0}$ have a small set C of order 1, i.e. $C \in \mathcal{E}^+$ and the minorization

$$(3.1) \qquad P \geq b I_C \otimes \nu$$

holds for some $b > 0$ and some probability measure ν on $(E, \mathcal{E})$. Let $\xi \in WCLT$ with the g.c.f. $\{\sigma_f^2;\ f \in B^*\}$. According to the discussion following the proof of Lemma III-3.1, the g.c.f. $\{\sigma_f^2;\ f \in B^*\}$ generates the reproducing kernel Hilbert space $(H, \|\cdot\|_H)$. Hence the rate function defined by (1.4) has representation (1.9). Here is the main result in this section.

Theorem 3.1. *Let $\{X_n\}_{n\geq 0}$ be an ergodic Markov chain with a small set of order 1 and let $\xi \in WCLT$. Let $\{b_n\}_{n\geq 1}$ be a non-decreasing sequence of positive numbers satisfying (1.1) and assume*

$$(3.2) \qquad S_n/b_n \longrightarrow 0 \quad \text{in probability.}$$

Then, for any initial distribution μ and for any open subset G of B,

$$(3.3) \qquad \liminf_{n\to\infty} \frac{n}{b_n^2} \log P_\mu\left\{\frac{S_n}{b_n} \in G\right\} \geq -\inf_{x\in G} I(x).$$

Let us make some comment before the proof. Despite the fact that it may not exist (Example I-1.1) in general, the small set of order 1 does exist in many interesting cases. When the state space is countable, for example, the minorization (3.1) is fulfilled with $b = 1$, $C = \{x\}$ and $\nu = P(x, \cdot)$ for any fixed $x \in E$. Therefore, we have the following corollary:

Corollary 3.2. *Let $\{X_n\}_{n\geq 0}$ be an ergodic Markov chain with countable state space and let $\xi \in WCLT$. Assume (3.2) holds. Then, we have the lower bound (3.3) for any initial distribution μ and for any open subset G of B.*

LIMIT THEOREMS FOR FUNCTIONALS 135

Proof of Theorem 3.1. Let the minorization (3.1) be fixed and consider the split chain $\tilde{P}$ (in the classical sense) generated by (3.1). Let the regeneration times $\{\tau(k)\}_{k\geq 0}$ be defined as in (I-2.19). Since $m = 1$, the random integer $i(n)$ defined in (I-3.10) is given by

$$i(n) = \sum_{k=0}^{n-1} I_{\alpha^*}(\Phi_k) \quad n = 1, 2, \cdots.$$

Define

$$\xi_k = \sum_{j=\tau(k-1)+1}^{\tau(k)} \xi(X_j) \quad k = 1, 2, \cdots.$$

Let the open subset G of B and the initial distribution μ be fixed and let $(H, ||\cdot||_H)$ be the reproducing kernel Hilbert space generated by the g.c.f. $\{\sigma_f^2; \ f \in B^*\}$. By (1.9) we need only to show that for any $x_o \in G \cap H$

(3.4) $$\liminf_{n\to\infty} \frac{n}{b_n^2} \log P_\mu\left\{\frac{S_n}{b_n} \in G\right\} \geq -I(x_o).$$

According to Lemma 2.1 of [26], the set

$$L = \{\pi^*(\alpha^*)E\xi_1 f(\xi_1); \ f \in B^*\}$$

is dense in H with respect to the Hilbert norm $||\cdot||_H$. By the continuity of $I(x)$ on H, we need only prove that (3.4) holds for any $x_o \in G \cap L$. Now, let $x_o \in G$ and $f_o \in B^*$ be fixed, with

(3.5) $$x_o = \pi^*(\alpha^*)E\xi_1 f_o(\xi_1).$$

Let $0 < \lambda < \pi^*(\alpha^*)$ be fixed, and choose $r > 0$ such that

(3.6) $$r < \frac{1}{2}\left(1 - \frac{\lambda}{\pi^*(\alpha^*)}\right)$$

and

$$\{x \in B; \ ||x - x_o|| < 3r\} \subset G.$$

From the condition (3.2), for sufficiently large n we have

$$\min_{1 \le k \le n-1} P_\nu\{||S_k|| \le \epsilon b_n\} \ge 2^{-1/2}.$$

By the minimal inequality (I-3.11) (with m=1) in Theorem I-3.4, and the fact

$$\{i(n) \ge k+1\} = \{\tau(k) \le n-1\} \quad (k \ge 0),$$

we have

$$P_\mu\left\{\frac{S_n}{b_n} \in G\right\} \ge P_\mu\left\{\left\|\frac{S_n}{b_n} - x_o\right\| < 3r\right\}$$
$$\ge 2^{-1/2} \tilde{P}_{\mu^*}\left\{\min_{0 \le k < i(n)} \left\|\frac{S_{\tau(k)+1}}{b_n} - x_o\right\| < 2r\right\}$$
$$\ge 2^{-1/2} \tilde{P}_{\mu^*}\left\{\left\|\frac{S_{\tau([n\lambda])+1}}{b_n} - x_o\right\| < 2r, \ i(n) \ge [n\delta]+1\right\}$$
$$= 2^{-1/2} \tilde{P}_{\mu^*}\left\{\left\|\frac{S_{\tau([n\lambda])+1}}{b_n} - x_o\right\| < 2r, \ \tau([n\lambda]) \le n-1\right\}.$$

Take n sufficiently large so that

$$\tilde{P}_{\mu^*}\{||S_{\tau(0)+1}|| < rb_n, \ \tau(0) < rn-1\} \ge 2^{-1/2}.$$

Note that

$$\begin{cases} S_{\tau([n\lambda])+1} = S_{\tau(0)+1} + \sum_{k=1}^{[n\lambda]} \xi_k & (n \ge 0) \\ \tau([n\lambda]) = \tau(0) + \sum_{k=1}^{[n\lambda]} (\tau(k) - \tau(k-1)) & (n \ge 0). \end{cases}$$

By Corollary I-2.3, $\left\{(S_{\tau(0)+1}, \tau(0)); \ (\xi_k, \tau(k) - \tau(k-1)); \ (k \ge 1)\right\}$ are independent $B \times \mathbf{R}$-valued random variables. Furthermore, $\{(\xi_k, \tau(k)-\tau(k-$

1))$\}_{k\geq 1}$ is an i.i.d. sequence. By the independence,

$$\tilde{P}_{\mu^*}\left\{\left\|\frac{S_{\tau([n\lambda])+1}}{b_n} - x_0\right\| < 2r, \quad \tau([n\lambda]) \leq n-1\right\}$$

$$\geq \tilde{P}_{\mu^*}\left\{\left\|\frac{1}{b_n}\sum_{k=1}^{[n\lambda]}\xi_k - x_0\right\| < r, \quad \|S_{\tau(0)+1}\| < rb_n,\right.$$

$$\left.\sum_{k=1}^{[n\lambda]}(\tau(k) - \tau(k-1)) < (1-r)n, \quad \tau(0) < rn-1\right\}$$

$$\geq 2^{-1/2}P\left\{\left\|\frac{1}{b_n}\sum_{k=1}^{[n\lambda]}\xi_k - x_0\right\| < r, \quad \sum_{k=1}^{[n\lambda]}(\tau(k) - \tau(k-1)) < (1-r)n\right\}.$$

Therefore,

(3.7)
$$P_\mu\left\{\frac{S_n}{b_n} \in G\right\}$$
$$\geq \frac{1}{2}P\left\{\left\|\frac{1}{b_n}\sum_{k=1}^{[n\lambda]}\xi_k - x_0\right\| < r, \quad \sum_{k=1}^{[n\lambda]}(\tau(k) - \tau(k-1)) < (1-r)n\right\}$$

for suffciently large n.

By the argument used to establish (III-4.31) and (III-4.36) (with b_n instead of $\sqrt{2nL_2n}$) we can prove that

(3.8)
$$\sum_{k=1}^{[n\lambda]}\xi_k/b_n \longrightarrow 0 \quad \text{in probability.}$$

Let $N > 0$ sufficiently large so that

$$P\{\|\xi_1\| \leq N, \quad \tau(1) - \tau(0) \leq N\} > 0,$$

and define

$$D_k = \{\|\xi_k\| \leq N, \quad \tau(k) - \tau(k-1) \leq N\},$$

$$\eta_k = \xi_k I_{D_k} \quad \text{and} \quad \zeta_k = \xi_k I_{D_k^c} \quad (k \geq 1).$$

We now claim that

(3.9) $$\frac{1}{b_n}\sum_{k=1}^{[n\lambda]}(\eta_k - E\eta_k) \longrightarrow 0 \quad \text{in probability,}$$

(3.10) $$\frac{1}{b_n}\sum_{k=1}^{[n\lambda]}(\zeta_k - E\zeta_k) \longrightarrow 0 \quad \text{in probability.}$$

Let the random sequence $\left\{\left(\xi_k', \, (\tau(k)-\tau(k-1))'\right)\right\}_{k\geq 1}$ be an independent copy of $\left\{\left(\xi_k, \, \tau(k)-\tau(k-1)\right)\right\}_{k\geq 1}$ and define

$$D_k' = \{\|\xi_k'\| \leq N, \, (\tau(k)-\tau(k))' \leq N\} \quad \text{and} \quad \eta_k' = \xi_k' I_{D_k'}.$$

From (3.8),

(3.11) $$\frac{1}{b_n}\sum_{k=1}^{[n\lambda]}(\xi_k - \xi_k') \longrightarrow 0 \quad \text{in probability.}$$

For each $k \geq 1$, it is easy to see that the random variable $(\xi_k - \xi_k')^*$ given by

$$(\xi_k - \xi_k')^* = (\xi_k - \xi_k')I_{D_k \cap D_k'} - (\xi_k - \xi_k')I_{(D_k \cap D_k')^c}$$

has the same distribution as $\xi_k - \xi_k'$. Note that

$$(\xi_k - \xi_k')I_{D_k \cap D_k'} = \frac{1}{2}\Big((\xi_k - \xi_k') + (\xi_k - \xi_k')^*\Big).$$

Hence, for each $t > 0$,

$$P\Big\{\Big\|\sum_{k=1}^{[n\lambda]}(\xi_k - \xi_k')I_{D_k \cap D_k'}\Big\| \geq t\Big\}$$

$$\leq P\Big\{\Big\|\sum_{k=1}^{[n\lambda]}(\xi_k - \xi_k')\Big\| \geq t\Big\} + P\Big\{\Big\|\sum_{k=1}^{[n\lambda]}(\xi_k - \xi_k')^*\Big\| \geq t\Big\}$$

$$= 2P\Big\{\Big\|\sum_{k=1}^{[n\lambda]}(\xi_k - \xi_k')\Big\| \geq t\Big\}.$$

Hence, from (3.11)

$$
(3.12) \qquad \frac{1}{b_n} \sum_{k=1}^{[n\lambda]} (\xi_k - \xi'_k) I_{D_k \cap D'_k} \longrightarrow 0 \quad \text{in probability.}
$$

By Lemma 7.2 in Ledoux-Talagrand (1991),

$$
(3.13) \qquad \lim_{n \to \infty} \frac{1}{b_n} E \Big\| \sum_{k=1}^{[n\lambda]} (\xi_k - \xi'_k) I_{D_k \cap D'_k} \Big\| = 0.
$$

Observe that

$$
(\xi_k - \xi'_k) I_{D_k \cap D'_k} = \eta_k I_{D'_k} - \eta'_k I_{D_k} \qquad k = 1, 2, \cdots
$$

and notice the independence between $\{(\eta_k, D_k)\}_{k \geq 1}$ and $\{(\eta'_k, D'_k)\}_{k \geq 1}$. Applying the Fubini's theorem and Jensen's inequality gives

$$
E \Big\| \sum_{k=1}^{[n\lambda]} (\xi_k - \xi'_k) I_{D_k \cap D'_k} \Big\| \geq E \Big\| \sum_{k=1}^{[n\lambda]} \big(\eta_k P(D_k) - E(\eta_k) I_{D_k} \big) \Big\|
$$

$$
\geq P(D_1) E \Big\| \sum_{k=1}^{[n\lambda]} \big(\eta_k - E(\eta_k) \big) \Big\| - \|E\eta_1\| \cdot E \Big| \sum_{k=1}^{[n\lambda]} \big(I_{D_k} - P(D_K) \big) \Big|.
$$

By the independence of $\{D_k\}_{k \geq 0}$ one can see that

$$
\lim_{n \to \infty} \frac{1}{b_n} E \Big| \sum_{k=1}^{[n\lambda]} \big(I_{D_k} - P(D_K) \big) \Big| = 0.
$$

From (3.13) one has that

$$
\lim_{n \to \infty} \frac{1}{b_n} E \Big\| \sum_{k=1}^{[n\lambda]} \big(\eta_k - E(\eta_k) \big) \Big\| = 0,
$$

which implies (3.9). Clearly, (3.10) follows from (3.8) and (3.9).

In the rest of the proof, we shall use the Lévy decomposition. Define the Bernoulli random variables α and β such that

$$P\{\alpha = 0\} = P\{\beta = 0\} = 1 - P(D_1) \quad \text{and} \quad P\{\alpha = 1\} = P\{\beta = 1\} = P(D_1)$$

and define the $B \times \mathbf{R}$-valued random variables (U, T) and (V, T') as follows:

$$\mathcal{L}\big((U, T)\big)(\cdot) = P(D_1 \cap \{(\xi_1, \tau(1) - \tau(0)) \in \cdot\})/P(D_1),$$

$$\mathcal{L}\big((V, T')\big)(\cdot) = P(D_1^c \cap \{(\xi_1, \tau(1) - \tau(0)) \in \cdot\})/P(D_1^c).$$

Further, we make α, β, (U, T) and (V, T') independent. Then,

$$\mathcal{L}\big((\xi_1, \ \tau(1) - \tau(0))\big) = \mathcal{L}\big(\beta(U, T) + (1 - \alpha)(V, T') + (\alpha - \beta)(U, T)\big),$$

$$\mathcal{L}\Big((\xi_1, \ \tau(1) - \tau(0))I_{D_1}\Big) = \mathcal{L}\big(\beta(U, T)\big),$$

$$\mathcal{L}\Big((\xi_1, \ \tau(1) - \tau(0))I_{D_1^c}\Big) = \mathcal{L}\big((1 - \alpha)(V, T')\big).$$

Now, let $\{\alpha_k\}_{k \geq 1}$, $\{\beta_k\}_{k \geq 1}$, $\{(U_k, T_k)\}_{k \geq 1}$ and $\{(V_k, T_k')\}_{k \geq 1}$ be the independent copies of α, β, (U, T) and (V, T'), respectively. We also make

$$\{\{\alpha_k\}_{k \geq 1}, \ \{\beta_k\}_{k \geq 1}, \ \{(U_k, T_k)\}_{k \geq 1}, \ \{(V_k, T_k)\}_{k \geq 1}\}$$

be an independent system. By (3.9),

(3.14) $$\frac{1}{b_n} \sum_{k=1}^{[n\lambda]} (\alpha_k - \beta_k) U_k \longrightarrow 0 \quad \text{in probability.}$$

From the independence between $\{\beta_k(U_k, T_k)\}_{k \geq 1}$ and $\{(1 - \alpha_k)(V_k, T_k')\}_{k \geq 1}$

and from (3.7),

$$P_\mu\left\{\frac{S_n}{b_n} \in G\right\} + P\left\{\|\sum_{k=1}^{[n\lambda]}(\alpha_k - \beta_k)U_k\| \geq \frac{r}{3}b_n\right\}$$

$$+ P\left\{|\sum_{k=1}^{[n\lambda]}(\alpha_k - \beta_k)T_k| \geq \frac{r}{3}n\right\}$$

$$\geq \frac{1}{2}P\left\{\|\frac{1}{b_n}\sum_{k=1}^{[n\lambda]}(\beta_k U_k + (1-\alpha_k)V_k) - x_o\| < \frac{2}{3}r,\right.$$

$$\left.\sum_{k=1}^{[n\lambda]}(\beta_k T_k + (1-\alpha_k)T'_k) < (1-\frac{4r}{3})n\right\}$$

(3.15)
$$\geq \frac{1}{2}P\left\{\|\frac{1}{b_n}\sum_{k=1}^{[n\lambda]}(\beta_k U_k - E(\beta_k U_k)) - x_o\| < \frac{r}{3},\right.$$

$$\left.\sum_{k=1}^{[n\lambda]}\beta_k T_k < (1-\frac{5r}{3})n\right\}$$

$$\times P\left\{\|\sum_{k=1}^{[n\lambda]}((1-\alpha_k)V_k - E((1-\alpha_k)V_k))\| < \frac{r}{3}b_n,\right.$$

$$\left.\sum_{k=1}^{[n\lambda]}(1-\alpha_k)T'_k < \frac{r}{3}n\right\}.$$

Take $N > 0$ sufficiently large so that

$$E((1-\alpha)T') = E(\tau(1) - \tau(0))I_{D_1^c} < (3\lambda)^{-1}r.$$

By the law of the large numbers for i.i.d. sequences

$$P\left\{\sum_{k=1}^{[n\lambda]}(1-\alpha_k)T'_k < \frac{r}{3}n\right\} \longrightarrow 1 \quad (n \to \infty).$$

From (3.10),

$$P\left\{\|\sum_{k=1}^{[n\lambda]}((1-\alpha_k)V_k - E((1-\alpha_k)V_k))\| < \frac{r}{3}b_n\right\} \longrightarrow 1 \quad (n \to \infty).$$

Therefore

$$P\Big\{||\sum_{k=1}^{[n\lambda]}((1-\alpha_k)V_k - E((1-\alpha_k)V_k))|| < \frac{r}{3}b_n,$$

(3.16)
$$\sum_{k=1}^{[n\lambda]}(1-\alpha_k)T'_k < \frac{r}{3}n\Big\} \longrightarrow 1 \quad (n\to\infty).$$

Since $(\alpha-\beta)T$ is bounded and centered, by the upper bounds in Cramér's large deviation result (see, e.g. Theorem 2.2.3 in [20]) there exists $u_1 > 0$ such that

(3.17)
$$P\Big\{|\sum_{k=1}^{[n\lambda]}(\alpha_k-\beta_k)T_k| \geq \frac{r}{3}n\Big\} \leq e^{-u_1 n}$$

for sufficiently large n. From (3.6),

$$E(\beta T) = E\big(\tau(1)-\tau(0)\big)I_{D_1} \leq E\big(\tau(1)-\tau(0)\big) = \pi^*(\alpha^*)^{-1} < \lambda^{-1}(1-\frac{5r}{3}),$$

where the second equality folows from Theorem 10.4.9 in Meyn-Tweedie (1993). Similarly, there exists $u_2 > 0$ such that

(3.18)
$$P\Big\{\sum_{k=1}^{[n\lambda]}\beta_k T_k \geq (1-\frac{5r}{3})n\Big\} \leq e^{-u_2 n}.$$

Since

$$P\Big\{||\frac{1}{b_n}\sum_{k=1}^{[n\lambda]}(\beta_k U_k - E(\beta_k U_k)) - x_o|| < \frac{r}{3}, \ \sum_{k=1}^{[n\lambda]}\beta_k T_k < (1-\frac{5r}{3})n\Big\}$$

$$+ P\Big\{\sum_{k=1}^{[n\lambda]}\beta_k T_k \geq (1-\frac{5r}{3})n\Big\}$$

$$\geq P\Big\{||\frac{1}{b_n}\sum_{k=1}^{[n\lambda]}(\beta_k U_k - E(\beta_k U_k)) - x_o|| < \frac{r}{3}\Big\}$$

$$= P\Big\{||\frac{1}{b_n}\sum_{k=1}^{[n\lambda]}(\eta_k - E(\eta_k)) - x_o|| < \frac{r}{3}\Big\},$$

from (3.18),

$$\liminf_{n\to\infty} \frac{n}{b_n^2} \log P\Big\{ \Big\| \frac{1}{b_n} \sum_{k=1}^{[n\lambda]} \big(\beta_k U_k - E(\beta_k U_k)\big) - x_o \Big\| < \frac{r}{3},$$

(3.19)
$$\sum_{k=1}^{[n\lambda]} \beta_k T_k < (1 - \frac{5r}{3})n \Big\}$$

$$\geq \liminf_{n\to\infty} \frac{n}{b_n^2} \log P\Big\{ \Big\| \frac{1}{b_n} \sum_{k=1}^{[n\lambda]} \big(\eta_k - E(\eta_k)\big) - x_o \Big\| < \frac{r}{3} \Big\}.$$

Combining (3.15), (3.16), (3.17) and (3.19) gives

(3.20)
$$\max\Big\{ \liminf_{n\to\infty} \frac{n}{b_n^2} \log P_\mu\Big\{ \frac{S_n}{b_n} \in G \Big\},$$
$$\limsup_{n\to\infty} \frac{n}{b_n^2} \log P\Big\{ \Big\| \sum_{k=1}^{[n\lambda]} (\alpha_k - \beta_k) U_k \Big\| \geq \frac{r}{3} b_n \Big\} \Big\}$$
$$\geq \liminf_{n\to\infty} \frac{n}{b_n^2} \log P\Big\{ \Big\| \frac{1}{b_n} \sum_{k=1}^{[n\lambda]} \big(\eta_k - E(\eta_k)\big) - x_o \Big\| < \frac{r}{3} \Big\}.$$

Now we recall some results obtained by the author ([11], [12]) on the moderate deviations for i.i.d. sequences, which states that if an i.i.d. B-valued random sequence $\{Z_n\}_{n\geq 1}$ satisfies

$$Ef(Z_1) = 0 \quad \text{and} \quad Ef^2(Z_1) < +\infty \quad \forall f \in B^*$$

and

$$\sum_{k=1}^{n} Z_k/b_n \longrightarrow 0 \quad \text{in probability,}$$

then (Theorem 1, [11]),

(3.21)
$$\liminf_{n\to\infty} \frac{n}{b_n^2} P\Big\{ \sum_{k=1}^{n} Z_k/b_n \in G \Big\} \geq -\inf_{x\in G} J(x)$$

for all open subsets G of B. If $\{Z_n\}_{n\geq 1}$ also satisfies

$$E\exp\{\epsilon\|Z_1\|\} < +\infty \quad \text{for some } \epsilon > 0,$$

then (Theorem 1, [12])

(3.22) $$\limsup_{n\to\infty} \frac{n}{b_n^2} P\left\{\sum_{k=1}^n Z_k/b_n \in F\right\} \leq -\inf_{x\in F} J(x)$$

for all closed subset F of B, where

(3.23) $$J(x) = \begin{cases} \frac{1}{2}\|x\|_{\bar{H}}^2 & x \in \bar{H} \\ +\infty & x \notin \bar{H}. \end{cases}$$

Here $(\bar{H}, \|\cdot\|_{\bar{H}})$ is the reproducing kernel Hilbert space generated by the covariance structures of Z_1 (see Lemma 2.1 in Goodman, Kuelbs and Zinn (1981) for the construction of the reproducing kernel Hilbert space in the i.i.d. context, in which case our definition presented in Section III-3 coincides with theirs).

Let H_o, H_N and H'_N be the reproducing kernel Hilbert spaces generated by the covariance structures of ξ_1, $\eta_1 - E\eta_1$ and $(\alpha-\beta)U_1$, respectively. Define

$$J_o(x) = \begin{cases} \frac{1}{2}\|x\|_{H_o}^2 & x \in H_o \\ +\infty & x \notin H_o, \end{cases}$$

$$J_N(x) = \begin{cases} \frac{1}{2}\|x\|_{H_N}^2 & x \in H_N \\ +\infty & x \notin H_N, \end{cases}$$

$$J'_N(x) = \begin{cases} \frac{1}{2}\|x\|_{H'_N}^2 & x \in H'_N \\ +\infty & x \notin H'_N. \end{cases}$$

By Theorem II-2.2 (with $f\circ \xi$ instead of ξ for each $f \in B^*$) we have $Ef(\xi_1) = 0$, $Ef^2(\xi_1) < +\infty$ and $\sigma_f^2 = \pi^*(\alpha^*)Ef^2(\xi_1)$ for all $f \in B^*$. From the constructions of the reproducing kernel Hilbert spaces (Lemma 2.1 of [26]),

(3.24) $$I(x) = \pi^*(\alpha^*)^{-1} \cdot J_o(x) \quad x \in B.$$

In view of (3.14), applying the upper bound (3.22) to the symmetric and bounded i.i.d. sequence $\{(\alpha_k - \beta_k)U_k\}_{k\geq 1}$ with $F = \{x;\ \|x\| \geq r/3\}$ and $[n\lambda]$ instead of n we have

$$\limsup_{n\to\infty} \frac{n}{b_n^2} \log P\Big\{\Big\|\sum_{k=1}^{[n\lambda]}(\alpha_k - \beta_k)U_k\Big\| \geq \frac{r}{3}b_n\Big\}$$
$$\leq -\lambda^{-1} \inf_{\|x\|\geq r/3} J'_N(x) \leq -\frac{r^2}{18\lambda\sigma_N^2},$$

where $\sigma_N^2 = \sup_{f\in B_1^*} Ef^2((\alpha-\beta)U)$ and the last inequality follows from Lemma 2.1-(iii) of [26]. By the definitions of α, β and U,

$$\sigma_N^2 = E|\alpha - \beta|^2 \cdot \sup_{f\in B_1^*} Ef^2(U) \leq 2P(D_1^c) \cdot \sigma^2,$$

where

(3.25) $\qquad \sigma^2 \equiv \sup_{f\in B_1^*} Ef^2(\xi_1) < +\infty \qquad$ (Lemma 2.1-(i) of [26]).

Hence

(3.26) $\quad \displaystyle\limsup_{n\to\infty} \frac{n}{b_n^2} \log P\Big\{\Big\|\sum_{k=1}^{[n\lambda]}(\alpha_k - \beta_k)U_k\Big\| \geq \frac{r}{3}b_n\Big\} \leq -\frac{r^2}{36\lambda\sigma^2 P(D_1^c)}.$

In view of (3.9), applying the lower bound (3.21) to the i.i.d. sequence $\{\eta_k - E\eta_k\}_{k\geq 1}$ with $G = \{x;\ \|x - x_o\| < r/3\}$ and $[n\lambda]$ instead of n gives

$$\liminf_{n\to\infty} \frac{n}{b_n^2} \log P\Big\{\Big\|\frac{1}{b_n}\sum_{k=1}^{[n\lambda]}(\eta_k - E(\eta_k)) - x_o\Big\| < \frac{r}{3}\Big\}$$
$$\geq -\lambda^{-1} \inf_{\|x-x_o\|<r/3} J_N(x).$$

Let

$$x_N = \pi^*(\alpha^*)E(\eta_1 - E\eta_1)f_o(\eta_1 - E\eta_1) \qquad (N > 0).$$

From the constructions of the reproducing kernel Hilbert spaces (Lemma 2.1 of [26]),

(3.27)
$$J_N(x_N) = \frac{1}{2}\pi^*(\alpha^*)^2 E f_o^2(\xi_1 I_{D_\lambda} - E\xi_1 I_{D_\lambda})$$
$$\leq \frac{1}{2}\pi^*(\alpha^*)^2 E f_o^2(\xi_1) = J_o(x_o) \quad (\lambda > 0)$$

and, in view of (3.5),

(3.28)
$$\|x_N - x_o\| = \sup_{g \in B_1^*} |g(x_N - x_o)|$$
$$= \pi^*(\alpha^*) \sup_{g \in B_1^*} |Eg(\eta_1 - E\eta_1)f_o(\eta_1 - E\eta_1) - Eg(\xi_1)f_o(\xi_1)|$$
$$= \pi^*(\alpha^*) \sup_{g \in B_1^*} |E\{g(\eta_1)f_o(\eta_1) - g(\xi_1)f_o(\xi_1)\} - Eg(\eta_1)Ef_o(\eta_1)|$$
$$= \pi^*(\alpha^*) \sup_{g \in B_1^*} |E\{g(\xi_1)f_o(\xi_1)I_{D_1^c}\} - Eg(\xi_1 I_{D_1^c})Ef_o(\xi_1 I_{D_1^c})|$$
$$\leq 2\pi^*(\alpha^*)\sigma \{Ef_o^2(\xi_1)I_{\{\|\xi_1\|>N\} \cup \{|\tau(1)-\tau(0)|>N\}}\}^{1/2} \longrightarrow 0$$

as $N \to \infty$. Take $N > 0$ sufficiently large so that $\|x_N - x_o\| < r/3$. By (3.27),
$$\inf_{\|x-x_o\|<\frac{r}{3}} J_N(x) \leq J_N(x_N) \leq J_o(x_o).$$

For sufficiently large N, therefore,

(3.29) $$\liminf_{n\to\infty} \frac{n}{b_n^2} \log P\left\{\left\|\frac{1}{b_n}\sum_{k=1}^{[n\lambda]}(\eta_k - E(\eta_k)) - x_o\right\| < \frac{r}{3}\right\} \geq -\lambda^{-1}J_o(x_o).$$

Combining (3.20), (3.26) and (3.29) gives
$$\max\left\{\liminf_{n\to\infty}\frac{n}{b_n^2}\log P_\mu\left\{\frac{S_n}{b_n} \in G\right\}, -\frac{r^2}{36\lambda\sigma^2 P(D_1^c)}\right\} \geq -\lambda^{-1}J_o(x_o).$$

Letting $N \longrightarrow \infty$ (that leads to $P(D_1^c) \longrightarrow 0$) gives
$$\liminf_{n\to\infty}\frac{n}{b_n^2}\log P_\mu\left\{\frac{S_n}{b_n} \in G\right\} \geq -\lambda^{-1}J_o(x_o).$$

In view of (3.24), letting $\lambda \longrightarrow \pi^*(\alpha^*)^-$ we finally obtain (3.4). ∎

IV-4. Lower bounds — when ξ is bounded

Suppose now that the measurable map $\xi: E \longrightarrow B$ is bounded and let $\xi \in WCLT$ with the g.c.f. $\{\sigma_f^2;\ f \in B^*\}$. According to the discussion following the proof of Lemma III-3.1, the g.c.f. $\{\sigma_f^2;\ f \in B^*\}$ generates the reproducing kernel Hilbert space $(H, ||\cdot||_H)$. Hence the rate function defined by (1.4) has representation (1.9). Here is the main result in this section.

Theorem 4.1. *Let $\{X_n\}_{n\geq 0}$ be an ergodic Markov chain and suppose ξ is bounded such that $\xi \in WCLT$. Let $\{b_n\}_{n\geq 1}$ be a sequence of non-decreasing positive numbers satisfying (1.1) and assume*

(4.1) $$S_n/b_n \longrightarrow 0 \text{ in probability.}$$

Then, for any initial distribution μ and for any open subset G of B,

(4.2) $$\liminf_{n\to\infty} \frac{n}{b_n^2} \log P_\mu\left\{\frac{S_n}{b_n} \in G\right\} \geq -\inf_{x\in G} I(x).$$

Remark 4.2. According to Theorem II-4.1 (with $f \circ \xi$ instead of ξ for each $f \in B^*$), "$\xi \in WCLT$" holds automatically with the g.c.f. $\{\sigma_f^2;\ f \in B^*\}$ represented by (1.6), provided $\{X_n\}_{n\geq 0}$ is ergodic of degree 2.

Proof of Theorem 4.1. Let the small set C be fixed. By definition there are some $m \geq 1$, $b > 0$ and some probability measure ν on $(E, \mathcal{E})$ such that

(4.3) $$P^m \geq b I_C \otimes \nu.$$

Because of Theorem 3.1 we only consider the situation that $m > 1$. Consider the split chain $\tilde{P}$ generated by (4.3) and let the regeneration times $\{\tau(k)\}_{k \geq 0}$ be defined as in (I-2.19) and $i(n)$ as in (I-3.10). Recall that the 1-dependent B-valued random sequence $\{\xi_k\}_{k \geq 1}$ is defined by

$$(4.4) \qquad \xi_k = \sum_{j=m(\tau(k-1)+1)}^{m\tau(k)+m-1} \xi(X_j) \qquad k = 1, 2, \cdots.$$

By Corollary I-2.4 and Theorem II-2.3 (with $f \circ \xi$ instead of ξ for each $f \in B^*$),

$$(4.5) \qquad Ef(\xi_1) = 0, \text{ and } Ef^2(\xi_1) < +\infty \qquad \forall f \in B^*,$$

$$(4.6) \qquad \sigma_f^2 = m^{-1}\pi^*(\alpha^*)\{Ef^2(\xi_1) + 2Ef(\xi_1)f(\xi_2)\} \qquad \forall f \in B^*.$$

Let the open subset G of B and the initial distribution μ be fixed and let $(H, ||\cdot||_H)$ be the reproducing kernel Hilbert space generated by the g.c.f. $\{\sigma_f^2;\ f \in B^*\}$. Like the proof of Theorem 3.1, it is enough to show that

$$(4.7) \qquad \liminf_{n \to \infty} \frac{n}{b_n^2} \log P_\mu \left\{ \frac{S_n}{b_n} \in G \right\} \geq -I(x_o)$$

holds for arbitrary $x_o \in G$ of the form

$$(4.8) \qquad x_o = m^{-1}\pi^*(\alpha^*)\{E\xi_1 f_o(\xi) + E\xi_1 f_o(\xi_2) + Ef_o(\xi_1)\xi_2\},$$

where $f_o \in B^*$. Let $0 < \lambda < \pi^*(\alpha^*)$ be fixed, and choose $r > 0$ such that

$$(4.9) \qquad r < \frac{1}{2}\left(1 - \frac{\lambda}{\pi^*(\alpha^*)}\right)$$

and

$$(4.10) \qquad \{x \in B;\ ||x - x_o|| < 10r\} \subset G.$$

By assumption
$$M \equiv 2(m-1)\|\xi\|_\infty < +\infty.$$

Let
$$p_n = \left[\frac{rb_n}{M}\right] \text{ and } q_n = \left[\frac{\lambda n}{mp_n}\right] \quad n = 1, 2, \cdots.$$

For each $n \geq 1$ and $1 \leq k \leq p_n$, define

$$\xi_{n,k} = \sum_{j=m(\tau((k-1)q_n)+1)}^{m\tau(kq_n)} \xi(X_j) - E\left(\sum_{j=m(\tau((k-1)q_n)+1)}^{m\tau(kq_n)} \xi(X_j)\right),$$

$$R_{n,k} = \sum_{j=m((k-1)q_n)+1}^{m\tau((k-1)q_n)+m-1} \xi(X_j) - E\left(\sum_{j=m((k-1)q_n)+1}^{m\tau((k-1)q_n)+m-1} \xi(X_j)\right).$$

From (4.5),
$$E(S_{m\tau(p_n q_n)+1} - S_{m\tau(0)+1}) = E\left(\sum_{k=1}^{p_n q_n} \xi_k\right) = 0.$$

Therefore,

(4.11) $$S_{m\tau(p_n q_n)+1} = S_{m\tau(0)+1} + \sum_{k=1}^{p_n} \xi_{n,k} + \sum_{k=1}^{p_n} R_{n,k} \quad (n \geq 1),$$

(4.12) $$\xi_{n,1} = \sum_{k=1}^{q_n} \xi_k - R_{n,2} \quad (n \geq 1).$$

Note that

(4.13) $$\|R_{n,k}\| \leq M \quad (1 \leq k \leq p_n, \ n \geq 1).$$

By a straightforward computation we obtain

(4.14) $$\lim_{n\to\infty} \frac{1}{q_n} E f_o^2(\xi_{n,1}) = E f_o^2(\xi) + 2E f_o(\xi_1) f_o(\xi_2),$$

(4.15) $$\lim_{n\to\infty} \frac{1}{q_n} \sup_{f\in B_1^*} Ef^2(\xi_{n,1}) = \sup_{f\in B_1^*} \{Ef^2(\xi) + 2Ef(\xi_1)f(\xi_2)\}.$$

Taking (4.5) and (4.12) into account, the argument used in the proof of Theorem 7.6 in Nummelin (1984) implies that

(4.16) $f_o(\xi_{n,1}/\sqrt{n}) \longrightarrow N\left(0,\ Ef_o^2(\xi) + 2Ef_o(\xi_1)f_o(\xi_2)\right)$ in distribution.

By Theorem I-2.2, for each $n \geq 1$,

$$\left\{\left(S_{m\tau(0)+1}, \tau(0)\right);\ \left(\xi_{n,k},\ \tau(kq_n) - \tau((k-1)q_n)\right);\ (1 \leq k \leq p_n)\right\}$$

are independent $B \times \mathbf{R}$-valued random variables. Furthermore,

$$\left\{\left(\xi_{n,k},\ \tau(kq_n) - \tau((k-1)q_n)\right)\right\}_{1\leq k\leq p_n}$$

is an i.i.d. sequence, and (4.13) implies

$$\left\|\sum_{k=1}^{p_n} R_{n,k}\right\| \leq rb_n.$$

In view of (4.10) and (4.11), the argument used to establish (3.7) implies

(4.17) $$P_\mu\left\{\frac{S_n}{b_n} \in G\right\} \geq \frac{1}{2}P\left\{\left\|\frac{1}{b_n}\sum_{k=1}^{p_n}\xi_{n,k} - x_o\right\| < 7r,\ \sum_{k=1}^{p_n}\left(\tau(kq_n) - \tau((k-1)q_n)\right) < (1-r)\frac{n}{m}\right\}$$

for sufficiently large n.

By the argument used to establish (III-4.31) and (III-4.36) (with b_n instead of $\sqrt{2nL_2n}$) we can prove that

$$\left(S_{m\tau(p_nq_n)+1} - S_{m\tau(0)+1}\right)/b_n \longrightarrow 0 \text{ in probability}.$$

Note that $p_n \sim M^{-1} r b_n$ and $\{R_{n,k}\}_{1 \leq k \leq p_n}$ are i.i.d. random variables. Applying the law of the large numbers gives

$$\frac{1}{b_n} \sum_{k=1}^{p_n} R_{n,k} \longrightarrow 0 \quad \text{in probability.}$$

Hence, from (4.11) we obtain

(4.18) $$\frac{1}{b_n} \sum_{k=1}^{p_n} \xi_{n,k} \longrightarrow 0 \quad \text{in probability.}$$

Since $\|\xi_1\| \leq \big(m(\tau(1) - \tau(0)) + m - 1\big) \|\xi\|_\infty$, the positivity of $\{X_n\}_{n \geq 0}$ implies that $E\|\xi_1\| < +\infty$. Applying the law of the large numbers to (4.12) gives

$$\frac{1}{q_n} \xi_{n,1} \longrightarrow 0 \quad \text{in probability.}$$

Thus, there exists a sequence $\{\delta_n\}_{n \geq 1}$ of decreasing positive numbers such that $\delta_n \longrightarrow 0^+$ and

(4.19) $$P\{\|\xi_{n,1}\| \leq \delta_n q_n\} \longrightarrow 1 \quad (n \to \infty).$$

By (4.9), one can choose $u > 0$ such that

(4.20) $$(1 - 2r)^{-1} \lambda < u^{-1} < \pi^*(\alpha^*).$$

Define

$$D_{n,k} = \{\|\xi_{n,k}\| \leq \delta_n q_n, \ \tau(k) - \tau(k-1) \leq u q_n\},$$

$$\eta_{n,k} = \xi_{n,k} I_{D_{n,k}} - E \xi_{n,k} I_{D_{n,k}} \quad \text{and} \quad \zeta_k = \xi_{n,k} I_{D^c_{n,k}} - E \xi_{n,k} I_{D^c_{n,k}}.$$

Similar to (3.9) and (3,10), (4.18) implies

(4.21) $$\frac{1}{b_n} \sum_{k=1}^{p_n} \eta_{n,k} \longrightarrow 0 \quad \text{in probability,}$$

(4.22) $$\frac{1}{b_n}\sum_{k=1}^{p_n}\zeta_{n,k}\longrightarrow 0 \text{ in probability.}$$

By the law of the large numbers for i.i.d. sequences,

(4.23) $$\lim_{n\to\infty}\frac{1}{q_n}\bigl(\tau(q_n)-\tau(0)\bigr)=E\bigl(\tau(1)-\tau(0)\bigr)=\pi^*(\alpha^*)^{-1} \text{ a.s.}$$

This, together with (4.19), implies that

(4.24) $$P(D_{n,1})\longrightarrow 1 \quad (n\to\infty).$$

By Theorem 5.4 in Billingsley (1968), (4.14) and (4.16) imply the uniform integrability of the random sequence $\{f_o^2(\xi_{n,1})/q_n\}_{n\geq 1}$. In particular, it follows from (4.14) and (4.24) that

(4.25) $$\lim_{n\to\infty}\frac{1}{q_n}Ef_o^2(\eta_{n,1})=Ef_o^2(\xi)+2Ef_o(\xi_1)f_o(\xi_2).$$

By the same reasoning we can prove the uniform integrability of $\{(\tau(q_n)-\tau(0))/q_n\}_{n\geq 1}$. In view of (4.24), we thus have

$$\lim_{n\to\infty}\frac{1}{q_n}E\bigl(\tau(q_n)-\tau(0)\bigr)I_{D_{n,1}^c}=0.$$

Consequently,

(4.26) $$\frac{1}{n}\sum_{k=1}^{p_n}\bigl(\tau(kq_n)-\tau((k-1)q_n)\bigr)I_{D_{n,k}^c}\longrightarrow 0 \text{ in probability.}$$

We proceed to the Lévy decomposition similarly as in Section IV-3. For each $n\geq 1$, define $\{\alpha_{n,k}\}_{1\leq k\leq p_n}$, $\{\beta_{n,k}\}_{1\leq k\leq p_n}$, $\{(U_{n,k},T_{n,k})\}_{1\leq k\leq p_n}$ and $\{(V_{n,k},T'_{n,k})\}_{1\leq k\leq p_n}$ to be four i.i.d. sequences with the following distributions, respectively:

$$P\{\alpha_{n,1}=0\}=P\{\beta_{n,1}=0\}=1-P(D_{n,1}),$$

$$P\{\alpha_{n,1}=1\}=P\{\beta_{n,1}=1\}=P(D_{n,1}),$$

$$\mathcal{L}\big((U_{n,1},T_{n,1})\big)(\cdot)=P(D_{n,1}\cap\{(\xi_{n,1},\tau(q_n)-\tau(0))\in\cdot\})/P(D_{n,1}),$$

$$\mathcal{L}\big((V_{n,1},T'_{n,1})\big)(\cdot)=P(D^c_{n,1}\cap\{(\xi_{n,1},\tau(q_n)-\tau(0))\in\cdot\})/P(D^c_{n,1}).$$

Then,

$$\mathcal{L}\big((\xi_{n,1},\ \tau(q_n)-\tau(0))\big)$$
$$=\mathcal{L}\big(\beta_{n,1}(U_{n,1},T_{n,1})+(1-\alpha_{n,1})(V_{n,1},T'_{n,1})+(\alpha_{n,1}-\beta_{n,1})(U_{n,1},T_{n,1})\big),$$

$$\mathcal{L}\Big((\xi_{n,1},\ \tau(q_n)-\tau(0))I_{D_{n,1}}\Big)=\mathcal{L}\big(\beta_{n,1}(U_{n,1},T_{n,1})\big),$$

$$\mathcal{L}\Big((\xi_{n,1},\ \tau(q_n)-\tau(0))I_{D^c_{n,1}}\Big)=\mathcal{L}\big((1-\alpha_{n,1})(V_{n,1},T'_{n,1})\big).$$

We also assume the independence beteween these sequences. From (4.17),

$$P_\mu\Big\{\frac{S_n}{b_n}\in G\Big\}$$
$$\geq\frac{1}{2}P\Big\{\Big\|\frac{1}{b_n}\sum_{k=1}^{p_n}(\beta_{n,k}U_{n,k}+(1-\alpha_{n,k})V_{n,k}$$
$$+(\alpha_{n,k}-\beta_{n,k})U_{n,k})-x_o\Big\|<7r,$$
$$\sum_{k=1}^{p_n}(\beta_{n,k}T_{n,k}+(1-\alpha_{n,k})T'_{n,k}+(\alpha_{n,k}-\beta_{n,k})T_{n,k})<(1-r)\frac{n}{m}\Big\}$$
$$\geq\frac{1}{2}P\Big\{\Big\|\frac{1}{b_n}\sum_{k=1}^{p_n}(\beta_{n,k}U_{n,k}+(1-\alpha_{n,k})V_{n,k})-x_o\Big\|<6r,$$
$$\sum_{k=1}^{p_n}(\alpha_{n,k}T_{n,k}+(1-\alpha_{n,k})T'^*_{n,k})<(1-r)\frac{n}{m}\Big\}$$
$$-\frac{1}{2}P\Big\{\Big\|\sum_{k=1}^{p_n}(\alpha_{n,k}-\beta_{n,k})U_{n,k}\Big\|\geq rb_n\Big\}.$$

In view of (4.20),

$$\sum_{k=1}^{p_n}\alpha_{n,k}T_{n,k}\stackrel{d}{=}\sum_{k=1}^{p_n}\big(\tau(kq_n)-\tau((k-1)q_n)\big)I_{D_{n,k}}\leq up_nq_n<(1-2r)\frac{n}{m}.$$

Therefore,

$$P_\mu\left\{\frac{S_n}{b_n} \in G\right\} + \frac{1}{2}P\left\{\|\sum_{k=1}^{p_n}(\alpha_{n,k} - \beta_{n,k})U_{n,k}\| \geq rb_n\right\}$$

$$\geq \frac{1}{2}P\Big\{\|\frac{1}{b_n}\sum_{k=1}^{p_n}\big(\beta_{n,k}U_{n,k} - E(\beta_{n,k}U_{n,k})\big) - x_o\| < 5r,$$

$$\|\sum_{k=1}^{p_n}\big((1-\alpha_{n,k})V_{n,k} - E((1-\alpha_{n,k})V_{n,k})\big)\| < rb_n,$$

$$\sum_{k=1}^{p_n}(1-\alpha_{n,k})T'_{n,k} < r\frac{n}{m}\Big\}$$

$$= \frac{1}{2}P\left\{\|\frac{1}{b_n}\sum_{k=1}^{p_n}\eta_{n,k} - x_o\| < 5r\right\}$$

$$\times P\left\{\|\sum_{k=1}^{p_n}\zeta_{n,k}\| < rb_n, \sum_{k=1}^{p_n}\big(\tau(kq_n) - \tau((k-1)q_n)\big)I_{D^c_{n,k}} < r\frac{n}{m}\right\}.$$

Taking (4.22) and (4.26) into account we have

(4.27)
$$\max\Big\{\liminf_{n\to\infty}\frac{n}{b_n^2}\log P_\mu\left\{\frac{S_n}{b_n} \in G\right\},$$
$$\limsup_{n\to\infty}\frac{n}{b_n^2}\log P\left\{\|\sum_{k=1}^{p_n}(\alpha_{n,k} - \beta_{n,k})U_{n,k}\| \geq rb_n\right\}\Big\}$$
$$\geq \liminf_{n\to\infty}\frac{n}{b_n^2}\log P\left\{\|\frac{1}{b_n}\sum_{k=1}^{p_n}\eta_{n,k} - x_o\| < 5r\right\}.$$

We now claim that

(4.28) $$\limsup_{n\to\infty}\frac{n}{b_n^2}\log P\left\{\|\sum_{k=1}^{p_n}(\alpha_{n,k} - \beta_{n,k})U_{n,k}\| \geq rb_n\right\} = -\infty.$$

From (4.21) we have

$$\frac{1}{b_n}\sum_{k=1}^{p_n}(\alpha_{n,k} - \beta_{n,k})U_{n,k} \longrightarrow 0 \text{ in probability.}$$

Note that $\{(\alpha_{n,k} - \beta_{n,k})U_{n,k}\}_{1 \leq k \leq p_n}$ are symmetric bounded i.i.d. B-valued random variables with

(4.29) $\qquad ||(\alpha_{n,k} - \beta_{n,k})U_{n,k}|| \leq \delta_n q_n \quad$ a.s. $\quad (1 \leq k \leq p_n).$

By Proposition 6.8 of [31] we have

(4.30) $\qquad \lim_{n \to \infty} \frac{1}{b_n} E||\sum_{k=1}^{p_n}(\alpha_{n,k} - \beta_{n,k})U_{n,k}|| = 0.$

Given $a > 0$, by Theorem 6.17 of [31],

$$P\{||\sum_{k=1}^{p_n}(\alpha_{n,k} - \beta_{n,k})U_{n,k}|| \geq 8e^a K_o E||\sum_{k=1}^{p_n}(\alpha_{n,k} - \beta_{n,k})U_{n,k}|| + \frac{3}{4}rb_n\}$$

$$\leq \exp\{-a\left[\frac{b_n^2}{n}\right]\} + P\{\sum_{k=1}^{[b_n^2/n]}||(\alpha_{n,k} - \beta_{n,k})U_{n,k}||^* \geq \frac{r}{4}b_n\}$$

$$+ 2\exp\{-\frac{r^2 b_n^2}{2468 e^a K_o p_n \sigma_n^2}\},$$

where K_o is the constant appearing in that theorem, the random sequence

$$\{||(\alpha_{n,k} - \beta_{n,k})U_{n,k}||^*\}_{1 \leq k \leq p_n}$$

is the non-increasing rearrangement of $\{||(\alpha_{n,k} - \beta_{n,k})U_{n,k}||\}_{1 \leq k \leq p_n}$ and by (4.15),

$$\sigma_n^2 \equiv \sup_{f \in B^*} E(\alpha_{n,1} - \beta_{n,1})^2 f^2(U_{n,1}) \leq 2P(D_{n,1}^c) \sup_{f \in B^*} Ef^2(\xi_{n,1})$$

$$\sim 2P(D_{n,1}^c)q_n \sup_{f \in B^*} \{Ef^2(\xi) + 2Ef(\xi_1)f(\xi_2)\} \quad (n \to \infty).$$

From (4.29) we can see that

$$\sum_{k=1}^{[b_n^2/n]} ||(\alpha_{n,k} - \beta_{n,k})U_{n,k}||^* < \frac{r}{4}b_n \quad \text{a.s.}$$

for sufficiently large n. Taking (4.24) and (4.30) into account, therefore, we have

$$\limsup_{n\to\infty} \frac{n}{b_n^2} \log P\Big\{\Big\|\sum_{k=1}^{p_n}(\alpha_{n,k}-\beta_{n,k})U_{n,k}\Big\| \geq rb_n\Big\} \leq -a.$$

Letting $a \longrightarrow +\infty$ gives (4.28).

Let

(4.31) $$x_n = \frac{\pi^*(\alpha^*)}{m} \cdot \frac{1}{q_n} E\eta_{n,1} f_o(\eta_{n,1}) \quad n=1,2,\cdots.$$

Then,

$$\|x_n - x_o\| = \sup_{g\in B_1^*} |g(x_n - x_o)|$$

$$= q_n^{-1} m^{-1} \pi^*(\alpha^*) \sup_{g\in B_1^*} |Eg(\eta_{n,1})f_o(\eta_{n,1}) - Eg(\xi_{n,1})f_o(\xi_{n,1})|$$

$$+ q_n^{-1} m^{-1} \pi^*(\alpha^*) \sup_{g\in B_1^*} \Big|Eg(\xi_{n,1})f_o(\xi_{n,1}) - Eg\Big(\sum_{k=1}^{q_n}\xi_k\Big)f_o\Big(\sum_{k=1}^{q_n}\xi_k\Big)\Big|$$

$$+ \sup_{g\in B_1^*} \Big|q_n^{-1} m^{-1}\pi^*(\alpha^*) Eg\Big(\sum_{k=1}^{q_n}\xi_k\Big)f_o\Big(\sum_{k=1}^{q_n}\xi_k\Big) - g(x_o)\Big|$$

$$= (I) + (II) + (III) \quad \text{(say)}.$$

Notice (4.15) and (4.24). Taking into account the fact that $\{f_o^2(\xi_{n,1})/q_n\}_{n\geq 1}$ is uniform integrable, the argument used in (3.28) implies that $(I) \longrightarrow 0$ as $n \to \infty$. From (4.12) and (4.13) it follows that $(II) \longrightarrow 0$ as $n \to \infty$. Note that

$$q_n^{-1} m^{-1} \pi^*(\alpha^*) Eg\Big(\sum_{k=1}^{q_n}\xi_k\Big) f_o\Big(\sum_{k=1}^{q_n}\xi_k\Big)$$

$$= m^{-1}\pi^*(\alpha^*)\Big\{Eg(\xi_1)f_o(\xi_1)$$

$$+ \frac{q_n-1}{q_n} Eg(\xi_1)f_o(\xi_2) + \frac{q_n-1}{q_n} Ef_o(\xi_1)g(\xi_2)\Big\}.$$

In view of (4.8), we have that $(III) \longrightarrow 0$ as $n \to \infty$. Therefore,

(4.32) $$\|x_n - x_o\| \longrightarrow 0 \quad (n\to\infty).$$

For each $n \geq 1$, define the psedometric d_n on B_1^* as follows:

$$d_n(f,g) = \frac{1}{q_n}\left\{E(f-g)^2(\eta_{n,1})\right\}^{1/2} \quad f,g \in B_1^*$$

and write $N(\epsilon, B_1^*, d_n)$, for each $n \geq 1$ and $\epsilon > 0$, to denote the minimal number of open balls of radius ϵ in the pseudometric d_n which are necessary to cover B_1^*, that is

$$N(\epsilon, B_1^*, d_n) = \min\{k;\ \text{there exist } f_1, \cdots, f_k \in B_1^* \text{ such that}$$
$$\min_{j \leq k} d_k(f_j, f) < \epsilon \text{ for all } f \in B_1^*\}.$$

Let $V_n(\epsilon)$ be the ϵ-net of the pseudometric space (B_1^*, d_n) such that

(4.33) $$\#(V_n(\epsilon)) = N(\delta, B_1^*, d_n)$$

and, let

$$U_n(\epsilon) = \left\{f \in B_1^*;\ Ef^2(\eta_{n,1}) \leq \frac{\epsilon^2}{4}q_n\right\}.$$

It is easy to see that for each $n \geq 1$ and $\epsilon > 0$,

(4.34) $$||x|| \leq ||f(x)||_{V_k(\epsilon)} + 2||f(x)||_{U_k(\epsilon)} \quad (x \in B),$$

where we write

$$||a_f||_U = \sup_{f \in U} |a_f|$$

for given family $\{a_f\}_{f \in U}$ of numbers indexed by a subset U of B_1^*. Taking into account (4.21) and the fact that for each $n \geq 1$

(4.35) $$||\eta_{n,k}|| \leq \delta_n q_n \quad \text{a.s.} \quad (1 \leq k \leq p_n),$$

the argument used to establish (3.15) in [13] (with $\{\eta_{n,k}\}$ instead of $\{Y_k\}$ and n/b_n^2 instead of t_n^2) implies that

(4.36) $$\lim_{n \to \infty} \frac{n}{b_n^2} \log N(\epsilon, B_1^*, d_n) = 0 \quad \forall \epsilon > 0.$$

Let $\{x_n\}$ be given in (4.31). By definition,

$$\|f(x_n)\|_{U_n(\epsilon)} = \frac{\pi^*(\alpha^*)}{m} \cdot \frac{1}{q_n} \sup_{f \in B_1^*} Ef(\eta_{n,1}) f_o(\eta_{n,1})$$

$$\leq \frac{\pi^*(\alpha^*)}{m} \cdot \frac{\epsilon}{2} \left\{ \frac{1}{q_n} Ef_o^2(\eta_{n,1}) \right\}^{1/2}.$$

In view of (4.14) and (4.32) we have

$$\|f(x_o)\|_{U_n(\epsilon)} \leq r$$

for suficiently large n and sufficiently small ϵ. By (4.34),

$$P\left\{ \left\| \frac{1}{b_n} \sum_{k=1}^{p_n} \eta_{n,k} - x_o \right\| < 5r \right\} + P\left\{ \left\| \sum_{k=1}^{p_n} f(\eta_{n,k}) \right\|_{U_n(\epsilon)} \geq rb_n \right\}$$

$$\geq P\left\{ \left\| \frac{1}{b_n} \sum_{k=1}^{p_n} f(\eta_{n,k}) - f(x_o) \right\|_{V_n(\epsilon)} < r \right\}.$$

This, together with (4.27) and (4.28), implies that

(4.37)
$$\max\left\{ \liminf_{n \to \infty} \frac{n}{b_n^2} \log P_\mu\left\{ \frac{S_n}{b_n} \in G \right\}, \right.$$
$$\left. \limsup_{n \to \infty} \frac{n}{b_n^2} \log P\left\{ \left\| \sum_{k=1}^{p_n} f(\eta_{n,k}) \right\|_{U_n(\epsilon)} \geq rb_n \right\} \right\}$$
$$\geq \liminf_{n \to \infty} \frac{n}{b_n^2} \log P\left\{ \left\| \frac{1}{b_n} \sum_{k=1}^{p_n} f(\eta_{n,k}) - f(x_o) \right\|_{V_n(\epsilon)} < r \right\}$$

for sufficiently small ϵ.

From (4.21) and (4.35), a standard way of symmetrization (Lemma 2.5, [25]) and the argument used to establish (4.28) imply that for any given $a > 0$,

(4.38) $$\limsup_{n \to \infty} \frac{n}{b_n^2} \log P\left\{ \left\| \sum_{k=1}^{p_n} f(\eta_{n,k}) \right\|_{U_n(\epsilon)} \geq rb_n \right\} \leq -a$$

for sufficiently small ϵ.

Let $\rho > 0$ be fixed but arbitrary and define, for each $n \geq 1$,

$$E_n = \Big\{(z_1, \cdots, z_{p_n}); \; z_k \in B \; (1 \leq k \leq p_n)$$
$$\text{and } \Big\|\frac{1}{b_n}\sum_{k=1}^{p_n} f(z_k) - f(x_o)\Big\|_{V_n(\epsilon)} < r\Big\},$$

$$F_n = \Big\{(z_1, \cdots, z_{p_n}); \; z_k \in B \; (1 \leq k \leq p_n)$$
$$\text{and } \frac{1}{b_n}\sum_{k=1}^{p_n} f(z_k) < f(x_o) + \rho\Big\}.$$

Then,

$$P\Big\{\Big\|\frac{1}{b_n}\sum_{k=1}^{p_n} f(\eta_{n,k}) - f(x_o)\Big\|_{V_n(\epsilon)} < r\Big\}$$

(4.39)
$$\geq \exp\Big\{-\frac{\pi^*(\alpha^*)b_n^2}{mp_n q_n}(f_o(x_o) + \rho)\Big\}$$
$$\times \int_{E_n \cap F_n} \exp\Big\{\frac{\pi^*(\alpha^*)b_n}{mp_n q_n}\sum_{k=1}^{p_n} f(z_k)\Big\}\mu_n(dz_1)\cdots\mu_n(z_{p_n}),$$

where

$$\mu_n = \mathcal{L}(\eta_{n,1}) \quad (n \geq 1).$$

If we let $\{Z_{n,k}; \; 1 \leq k \leq p_n \; n \geq 1\}$ be a triagular array of rowwise i.i.d. B-valued random variables with the distributions

$$\mathcal{L}(Z_{n,k})$$
$$= \int_{(\cdot)} \exp\Big\{\frac{\pi^*(\alpha^*)b_n}{mp_n q_n}f_o(z)\Big\}\mu_n(dz) \Big/ \int \exp\Big\{\frac{\pi^*(\alpha^*)b_n}{mp_n q_n}f_o(z)\Big\}\mu_n(dz),$$

then (4.39) can be rewritten as

$$P\Big\{\Big\|\frac{1}{b_n}\sum_{k=1}^{p_n} f(\eta_{n,k}) - f(x_o)\Big\|_{V_n(\epsilon)} < r\Big\}$$

(4.40)
$$\geq \exp\Big\{-\frac{\pi^*(\alpha^*)b_n^2}{mp_n q_n}(f_o(x_o) + \rho)\Big\}\Big(E \exp\Big\{\frac{\pi^*(\alpha^*)b_n}{mp_n q_n}f_o(\eta_{n,1})\Big\}\Big)^{p_n}$$
$$\times P\{(Z_1, \cdots, Z_{p_n}) \in E_n \cap F_n\}.$$

We now claim that

(4.41) $$P\{(Z_1,\cdots,Z_{p_n}) \in E_n \cap F_n\} \longrightarrow 1 \quad (n \to \infty).$$

In order to have (4.41), it is sufficient to prove

(4.42) $$P\Big\{\frac{1}{b_n}\sum_{k=1}^{p_n} f(Z_{n,k}) \geq f(x_o) + \rho\Big\} \longrightarrow 0 \quad (n \to \infty),$$

(4.43) $$P\Big\{\Big\|\frac{1}{b_n}\sum_{k=1}^{p_n} f(Z_{n,k}) - f(x_o)\Big\|_{V_n(\epsilon)} \geq r\Big\} \longrightarrow 0 \quad (n \to \infty).$$

Let $\{x_n\}$ be defined in (4.31) and let

$$\lambda_n = E\exp\Big\{\frac{\pi^*(\alpha^*)b_n^2}{mp_n q_n} f_o(\eta_{n,1})\Big\} \quad n = 1,2,\cdots.$$

Then, $\lambda_n \longrightarrow 1$ as $n \to \infty$ and,

$$\Big\|\frac{1}{b_n}\sum_{k=1}^{p_n} EZ_{n,k} - \frac{1}{\lambda_n}x_n\Big\| = \sup_{g \in B_1^*}\Big|\frac{p_n}{b_n}Eg(Z_{n,1}) - \frac{1}{\lambda_n}g(x_n)\Big|$$

$$= \frac{1}{\lambda_n}\sup_{g \in B_1^*}\Big|\frac{p_n}{b_n}Eg(\eta_{n,1})\exp\Big\{\frac{\pi^*(\alpha^*)b_n}{mp_n q_n}f_o(\eta_{n,1})\Big\}$$

$$- \frac{\pi^*(\alpha^*)}{mq_n}Eg(\eta_{n,1})f_o(\eta_{n,1})\Big|$$

$$= \frac{1}{\lambda_n}\cdot\frac{p_n}{b_n}\sup_{g \in B_1^*}\Big|Eg(\eta_{n,1})\Big(\exp\Big\{\frac{\pi^*(\alpha^*)b_n}{mp_n q_n}f_o(\eta_{n,1})\Big\} - 1$$

$$- \frac{\pi^*(\alpha^*)b_n}{mp_n q_n}f_o(\eta_{n,1})\Big)\Big|$$

$$\longrightarrow 0$$

as $n \to \infty$, where the last step follows from (4.15), (4.35) and the Taylor's expansion. This, together with (4.32), implies that

(4.44) $$\lim_{n \to \infty}\Big\|\frac{1}{b_n}\sum_{k=1}^{p_n} EZ_{n,k} - x_o\Big\| = 0.$$

From (4.15),

$$\sup_{f \in B_1^*} \sum_{k=1}^{p_n} Ef^2(Z_{n,k} - EZ_{n,k}) = p_n \sup_{f \in B_1^*} Ef^2(Z_{n,1} - EZ_{n,1})$$

(4.45)
$$\sim p_n \sup_{f \in B_1^*} Ef^2(\eta_{n,1}) \leq p_n \sup_{f \in B_1^*} Ef^2(\xi_{n,1})$$

$$\sim \frac{\lambda n}{m} \sup_{f \in B^*} \{Ef^2(\xi) + 2Ef(\xi_1)f(\xi_2)\} \quad (n \to \infty).$$

In particular,

$$\frac{1}{b_n} \sum_{k=1}^{p_n} f_o(Z_{n,k} - EZ_{n,k}) \longrightarrow 0 \quad \text{in probability}.$$

Hence, (4.42) follows from (4.44).

By (4.44), we have

$$\left\| \frac{1}{b_n} \sum_{k=1}^{p_n} EZ_{n,k} - x_o \right\| < \frac{r}{2}$$

for sufficiently large n. From (4.33) we have

$$P\left\{ \left\| \frac{1}{b_n} \sum_{k=1}^{p_n} f(Z_{n,k}) - f(x_o) \right\|_{V_n(\epsilon)} \geq r \right\}$$

(4.46)
$$\leq P\left\{ \left\| \sum_{k=1}^{p_n} f(Z_{n,k} - EZ_{n,k}) \right\|_{V_n(\epsilon)} \geq \frac{r}{2} b_n \right\}$$

$$\leq N(\epsilon, B_1^*, d_n) \sup_{f \in B_1^*} P\left\{ \left| \sum_{k=1}^{p_n} f(Z_{n,k} - EZ_{n,k}) \right| \geq \frac{r}{2} b_n \right\}.$$

Taking into account (4.45) and the fact that

$$\|Z_{n,1} - EZ_{n,1}\| \leq 2\delta_n q_n \quad \text{a.s.} \quad n = 1, 2, \cdots,$$

a standard exponential estimation due to Kolmogorov implies that there exists $u > 0$ such that

$$\sup_{f \in B_1^*} P\left\{ \left| \sum_{k=1}^{p_n} f(Z_{n,k} - EZ_{n,k}) \right| \geq \frac{r}{2} b_n \right\} \leq \exp\left\{ -u \frac{b_n^2}{n} \right\}$$

holds for sufficiently large n. Therefore, (4.43) follows from (4.36) and (4.46).

In view of (4.25) and (4.35), applying the Taylor's expansion gives

$$
(4.47) \quad \begin{aligned} & E \exp\left\{\frac{\pi^*(\alpha^*)b_n}{mp_n q_n} f_o(\eta_{n,1})\right\} \\ & = 1 + \frac{\pi^*(\alpha^*)^2 b_n^2}{2m\lambda n p_n}\left(E f_o^2(\xi) + 2E f_o(\xi_1) f_o(\xi_2)\right) + o\left(\frac{b_n^2}{np_n}\right) \quad (n\to\infty). \end{aligned}
$$

By (III-3.26), (1.9) and (4.8),

$$(4.48) \quad f_o(x_o) = m^{-1}\pi^*(\alpha^*)\big(E f_o^2(\xi) + 2E f_o(\xi_1) f_o(\xi_2)\big) = 2I(x_o).$$

Combining (4.40), (4.41), (4.47) and (4.48) gives

$$\liminf_{n\to\infty} \frac{n}{b_n^2} \log P\Big\{\Big\|\frac{1}{b_n}\sum_{k=1}^{p_n} f(\eta_{n,k}) - f(x_o)\Big\|_{V_n(\epsilon)} < r\Big\} \geq -\frac{\pi^*(\alpha^*)}{\lambda}\big(I(x_o) + \rho\big).$$

This, along with (4.37) and (4.38), gives that

$$\max\Big\{\liminf_{n\to\infty} \frac{n}{b_n^2} \log P_\mu\Big\{\frac{S_n}{b_n} \in G\Big\},\ -a\Big\} \geq -\frac{\pi^*(\alpha^*)}{\lambda}\big(I(x_o) + \rho\big).$$

Letting $a \longrightarrow +\infty$, $\rho \longrightarrow 0^+$ then $\lambda \longrightarrow \pi^*(\alpha^*)^-$ we obtain (4.7). ∎

IV-5. Lower bounds — when $\xi \in CLT$

In Chapter II, we discussed the one dimensional central limit theorem. With no difficulty, we can generalize this concept to the vector setting. Given an ergodic Markov chain $\{X_n\}_{n\geq 0}$, we say the measurable map $\xi\colon E \longrightarrow B$ satisfies the **central limit theorem** (write $\xi \in CLT$ for short) if there exists a centered Gaussian measure γ on B such that

$$(5.1) \quad S_n/\sqrt{n} \longrightarrow \gamma \text{ in distribution.}$$

Note that the CLT is a strengthened form of the WCLT defined in Section III-3. More precisely, if (5.1) holds then $\xi \in WCLT$ with the g.c.f. $\{\sigma_f^2;\ f \in B^*\}$ given by

$$(5.2) \qquad \sigma_f^2 = \int f^2(x)\gamma(dx) \quad f \in B^*.$$

Hence, the reproducing kernel Hilbert space $(H, \|\cdot\|_H)$ generated by the g.c.f. $\{\sigma_f^2;\ f \in B^*\}$ coincides with the one generated by covariance structure of γ in the Goodman-Kuelbs-Zinn (Lemma 2.1, [26]) sense. Our main result in this section states that the CLT implies the lower bounds of moderate deviations.

Theorem 5.1. *Let $\{X_n\}_{n\geq 0}$ be an ergodic Markov chain and assume that $\xi \in CLT$. Then, for any sequence $\{b_n\}_{n\geq 1}$ satisfying (1.1), any initial distribution μ and any open subset G of B,*

$$(5.3) \qquad \liminf_{n\to\infty} \frac{n}{b_n^2} \log P_\mu\left\{\frac{S_n}{b_n} \in G\right\} \geq -\inf_{x\in G} I(x).$$

Proof. Let C be a small set. By Theorem 5.2.3 in [33], there are some $m \geq 1$, $b > 0$ and some probability measure ν on $(E, \mathcal{E})$ such that $\nu(C) > 0$ and

$$(5.4) \qquad P^m \geq bI_C \otimes \nu.$$

We may assume that $\nu(C) = 1$, for otherwise we can take $\nu_C = \nu(C)^{-1}\nu(C \cap \cdot)$ instead of ν and $b\nu(C)$ instead of b in (5.4).

Let γ be the Gaussian measure on B satisfying (5.10). According to the large deviations for Gaussian measures (see, e.g. Theorem 3.4.5 of [21]),

$$\liminf_{t\to +\infty} \frac{1}{t^2} \log \gamma(tG) \geq -\inf_{x\in G} I(x)$$

for every open subset G of B. Fix the open set G in (5.3). We need only to prove, therefore, that for every $x_o \in G$ there exists $r > 0$ such that

$$(5.5) \qquad \liminf_{n \to \infty} \frac{n}{b_n^2} \log P_\mu \left\{ \frac{S_n}{b_n} \in G \right\} \geq \frac{1}{t^2} \log b\gamma(tB_{x_o, r})$$

holds for each $t > 0$, where

$$B_{x_o, r} = \{x \in B; \ ||x - x_o|| < r\}.$$

Let $x_o \in G$, $t > 0$ be fixed and choose $r > 0$ be such that

$$\{x \in B; \ ||x - x_o|| < 3r\} \subset G.$$

We may assume that $\gamma(tB_{x_o, r}) > 0$. Hence, one can choose $0 < \rho < 1/4$ such that

$$(5.6) \qquad b\gamma(tB_{x_o, r}) > \rho.$$

We may also assume that there exists a constant $M > 0$ such that

$$(5.7) \qquad C \subset \left\{ x; \ P_x \left\{ \sum_{j=1}^{m-1} ||\xi(X_j)|| > M \right\} \leq b\rho \right\}$$

for otherwise, we can take $M > 0$ sufficiently large so that the set C_M given by

$$C_M \equiv C \cap \left\{ x; \ P_x \left\{ \sum_{j=1}^{m-1} ||\xi(X_j)|| > M \right\} \leq \frac{b\rho}{2} \right\}$$

satisfies

$$\nu(C_M) > \frac{1}{2} \quad \text{and} \quad \nu(C_M) b\gamma(tB_{x_o, r}) > \rho,$$

and use C_M instead of C, $\nu_M = \nu(C_M)^{-1} \nu(C_M \cap \cdot)$ instead of ν and $\nu(C_M) b$ instead of b in (5.4) and, if we can prove (5.5) under the additional assumption (5.7) then,

$$\liminf_{n \to \infty} \frac{n}{b_n^2} \log P_\mu \left\{ \frac{S_n}{b_n} \in G \right\} \geq \frac{1}{t^2} \log \nu(C_M) b\gamma(tB_{x_o, r})$$

under such substitution, which leads to (5.5) as $M \longrightarrow +\infty$.

Unlike the previous proof, the argument used here does not depend on split method. Recall (Section I-3) that $\{i_C(n)\}_{n\geq 1}$ and $\{\tau_C^m(k)\}_{k\geq 0}$ are defined as follows:

$$i_C(n) = \sum_{k=1}^{[n/m]-1} I_C(X_{km}) \quad n = 1, 2, \cdots,$$

$$\begin{cases} \tau_C^m(1) = \inf\{n \geq 1;\ X_{mn} \in C\} \\ \tau_C^m(k+1) = \inf\{n > \tau_C^m(k);\ X_{mn} \in C\} \quad (k \geq 1). \end{cases}$$

Then, for each $k \geq 1$, $m\tau_C^m(k)$ is a stopping time w.r.t. $\{X_n\}_{n\geq 0}$. We first claim that

(5.8) $$\sqrt{\frac{\pi(C)}{m}} \cdot \frac{S_{m\tau_C^m(n)+1}}{\sqrt{n}} \longrightarrow \gamma \text{ in distribution.}$$

Let $a(n) = [mn\pi(C)^{-1}]$ $(n \geq 1)$. From (5.1) one can easily see that

$$\sqrt{\frac{\pi(C)}{m}} \cdot \frac{S_{a(n)}}{\sqrt{n}} \longrightarrow \gamma \text{ in distribution.}$$

To prove (5.8), therefore, it suffices to show that

(5.9) $$\frac{1}{\sqrt{n}}\left(S_{a(n)} - S_{m\tau_C^m(n)+1}\right) \longrightarrow 0 \text{ in probability.}$$

Because of ergodicity, we may take π as the initial distribution in the proof of (5.9).

Given $\epsilon > 0$, by (5.1) there exists $\delta > 0$ such that

(5.10) $$\max_{k \leq 3\delta n} P_\pi\{\|S_k\| \geq \frac{\epsilon}{3}\sqrt{n}\} < \frac{\epsilon}{3}b,$$

(5.11) $$\max_{k \leq 3\delta n} P_\nu\{\|S_k\| \geq \frac{\epsilon}{3}\sqrt{n}\} < \frac{1}{4}$$

holds for sufficiently large n. We may assume that $\delta < m\pi(C)^{-1}$. Take n sufficiently large such that $a(n) > \delta n$ and let

$$k(n) = \max\{k;\ km + 1 \le a(n) - \delta n\}.$$

Then, on the event $\{|m\tau_C^m(n) + 1 - a(n)| < \delta n\}$,

$$\|S_{a(n)} - S_{m\tau_C^m(n)+1}\| \le \|S_{a(n)} - S_{mk(n)}\| + \|S_{m\tau_C^m(n)+1} - S_{mk(n)}\|$$

$$\le \|S_{a(n)} - S_{mk(n)}\| + \max_{k \le i'_C([2(m+\delta n)])} \|S_{m\hat{\tau}_C^m(k)+1} - S_{mk(n)}\|,$$

where

$$i'_C(k) = \sum_{j=k(n)+1}^{k(n)+[k/m]-1} I_C(X_{jm}) \quad k = 1, 2, \cdots,$$

$$\begin{cases} \hat{\tau}_C^m(1) = \inf\{j \ge k(n);\ X_{mj} \in C\} \\ \hat{\tau}_C^m(k+1) = \inf\{j > \hat{\tau}_C^m(k);\ X_{mj} \in C\} \quad (k \ge 1) \end{cases}$$

and the last step follows from the fact that on the event $\{|m\tau_C^m(n) + 1 - a(n)| < \delta n\}$,

$$k(n) < \tau_C^m(n) \le k(n) + \left[\frac{[2(m+\delta n)]}{m}\right] - 1.$$

Note that when $n > 2\delta^{-1}m$,

$$\max_{k \le i'_C([2(m+\delta n)])} \|S_{m\hat{\tau}_C^m(k)+1} - S_{mk(n)}\|$$

$$\le \max_{k \le i'_C([3\delta n])} \|S_{m\hat{\tau}_C^m(k)+1} - S_{mk(n)}\| = \left(\max_{k \le i_C([3\delta n])} \|S_{m\tau_C^m(k)+1}\|\right) \circ \theta^{mk(n)}.$$

By stationarity,

$$P_\pi\{\|S_{a(n)} - S_{m\tau_C^m(n)+1}\| \ge \epsilon\}$$

(5.12) $\le P_\pi\{|m\tau_C^m(n) + 1 - a(n)| \ge \delta n\} + P_\pi\{\|S_{a(n)-mk(n)}\| \ge \frac{\epsilon}{3}\sqrt{n}\}$

$$+ P_\pi\{\max_{k \le i_C([3\delta n])} \|S_{m\tau_C^m(k)+1}\| \ge \frac{2\epsilon}{3}\sqrt{n}\}.$$

By the law of the large numbers for Markov chains (Theorem 7.17 of [33]),

$$\lim_{n\to\infty} \frac{1}{n} i_C(n) = \frac{\pi(C)}{m} \quad \text{a.s.}$$

Hence, from the relation

(5.13) $\quad \{i_C(n) \geq k\} = \{\tau_C^m(k) \leq [n/m] - 1\} \quad k = 1, 2, \cdots$

one can easily see that

(5.14) $\quad \displaystyle\lim_{n\to\infty} \frac{1}{n} \tau_C^m(n) = \frac{1}{\pi(C)} \quad \text{a.s.}$

In paticular,

(5.15) $\quad P_\pi\{|m\tau_C^m(n) + 1 - a(n)| \geq \delta n\} \longrightarrow 0 \quad (n \to \infty).$

From (5.10),

(5.16) $\quad P_\pi\{\|S_{a(n)-mk(n)}\| \geq \frac{\epsilon}{3}\sqrt{n}\} < \frac{\epsilon}{3}$

for sufficiently large n. Taking (5.10) and (5.11) into account, applying the maximal inequality in Theorem I-3.2 gives

(5.17) $\quad \begin{aligned} &P_\pi\{\max_{k\leq i_C([3\delta n])} \|S_{m\tau_C^m(k)+1}\| > \frac{2\epsilon}{3}\sqrt{n}\} \\ &\leq 2P_\pi\{\|S_{[3\delta n]}\| \geq \frac{\epsilon}{3}\sqrt{n}\} < \frac{2\epsilon}{3} \end{aligned}$

for sufficiently large n. Since ϵ is arbitrary, (5.9) follows from (5.12), (5.15), (5.16) and (5.17).

Let $0 < \lambda < \pi(C)$ be fixed but arbitrary. Define

$$p_n = \left[\frac{\lambda^2 n^2 t^2}{m\pi(C)b_n^2}\right], \quad q_n = \left[\frac{\lambda n}{mp_n}\right] \quad \text{and} \quad r_n = \sqrt{\frac{\pi(C)}{m}} \cdot \frac{b_n}{tq_n} \quad (n \geq 1).$$

From (5.14),
$$P_\nu\left\{\tau_C^m(p_n-1) < \frac{1}{q_n}\left[\frac{n}{m}\right]-1\right\} \longrightarrow 1 \quad (n\to\infty).$$

Note that $r_n \sim \sqrt{p_n-1}$ as $n\to\infty$. By (5.8),
$$\sqrt{\frac{\pi(C)}{m}} \cdot \frac{S_{m\tau_C^m(p_n-1)+1}}{r_n} \longrightarrow \gamma \text{ in distribution.}$$

In particular,

(5.18)
$$\liminf_{n\to\infty} P_\nu\left\{\left\|\sqrt{\frac{\pi(C)}{m}} \cdot \frac{S_{m\tau_C^m(p_n-1)+1}}{tr_n} - x_o\right\| < r,\right.$$
$$\left.\tau_C^m(p_n-1) < \frac{1}{q_n}\left[\frac{n}{m}\right]-1\right\} \geq \gamma(tB_{x_o,r}).$$

Hence, from (5.6) we have

(5.19)
$$bP_\nu\left\{\left\|\sqrt{\frac{\pi(C)}{m}} \cdot \frac{S_{m\tau_C^m(p_n-1)+1}}{tr_n} - x_o\right\| < r,\right.$$
$$\left.\tau_C^m(p_n-1) < \frac{1}{q_n}\left[\frac{n}{m}\right]-1\right\} > \rho$$

for sufficiently large n. Let n be sufficiently large such that
$$\min_{k\leq n} P\{\|S_k\| < \frac{r}{2}b_n\} > \frac{1}{2}.$$

In view of (5.7) and (5.13), applying the minimal inequality (I-3.4) in Theorem I-3.2 gives

(5.20)
$$P_\mu\left\{\frac{S_n}{b_n} \in G\right\} \geq P_\mu\left\{\left\|\frac{S_n}{b_n} - x_o\right\| < 3r\right\}$$
$$\geq b\left(\frac{1}{2}-\rho\right)P_\mu\left\{\min_{k\leq i_C(n)}\left\|\frac{S_{m\tau_C^m(k)+1}}{b_n} - x_o\right\| < 2r\right\}$$
$$\geq b\left(\frac{1}{2}-\rho\right)P_\mu\left\{\left\|\frac{S_{m\tau_C^m(q_np_n)+1}}{b_n} - x_o\right\| < 2r,\; i_C(n) \leq q_np_n\right\}$$
$$= b\left(\frac{1}{2}-\rho\right)P_\mu\left\{\left\|\frac{S_{m\tau_C^m(q_np_n)+1}}{b_n} - x_o\right\| < 2r,\; \tau_C^m(q_np_n) < \left[\frac{n}{m}\right]\right\}$$
$$= b\left(\frac{1}{2}-\rho\right)P_\mu\left\{\left\|\sqrt{\frac{\pi(C)}{m}} \cdot \frac{S_{m\tau_C^m(q_np_n)+1}}{tq_nr_n} - x_o\right\| < 2r,\right.$$
$$\left.\tau_C^m(q_np_n) < \left[\frac{n}{m}\right]\right\}.$$

Note that

$$\left\{\left\|\sqrt{\frac{\pi(C)}{m}} \cdot \frac{S_{m\tau_C^m(q_np_n)+1}}{tq_nr_n} - x_0\right\| < 2r,\ \tau_C^m(q_np_n) < \left[\frac{n}{m}\right]\right\}$$

$$\supset \left\{\left\|\sqrt{\frac{\pi(C)}{m}} \cdot \frac{S_{m\tau_C^m((q_n-1)p_n)+1}}{t(q_n-1)r_n} - x_0\right\| < 2r,\right.$$

$$\left.\tau_C^m((q_n-1)p_n) < \frac{q_n-1}{q_n}\left[\frac{n}{m}\right]\right\}$$

$$\cap \left\{\left\|\sqrt{\frac{\pi(C)}{m}} \cdot \frac{\left(S_{m\tau_C^m(q_np_n)+1} - S_{m\tau_C^m((q_n-1)p_n)+1}\right)}{tr_n} - x_0\right\| < 2r,\right.$$

$$\left.\tau_C^m(q_np_n) - \tau_C^m((q_n-1)p_n) < \frac{1}{q_n}\left[\frac{n}{m}\right]\right\}$$

$$= A_1 \cap A_2 \quad \text{(say)}.$$

Let n be sufficiently large so that $r_n > Mr^{-1}t^{-1}\sqrt{\pi(C)/m}$. By the Markov property,

$$P_\mu\left\{\left\|\sqrt{\frac{\pi(C)}{m}} \cdot \frac{S_{m\tau_C^m(q_np_n)+1}}{tq_nr_n} - x_0\right\| < 2r,\ \tau_C^m(q_np_n) < \left[\frac{n}{m}\right]\right\}$$

$$\geq E_\mu\left(I_{A_1} \cdot P_{X_{m\tau_C^m((q_n-1)p_n)}}\left\{\left\|\sqrt{\frac{\pi(C)}{m}} \cdot \frac{(S_{m\tau_C^m(p_n)+1} - S_1)}{tr_n}\right.\right.\right.$$

(5.21)
$$\left.\left.\left. - x_0\right\| < 2r,\ \tau_C^m(p_n) < \frac{1}{q_n}\left[\frac{n}{m}\right]\right\}\right)$$

$$\geq E_\mu\left(I_{A_1} \cdot P_{X_{m\tau_C^m((q_n-1)p_n)}}\left\{\left\|\sqrt{\frac{\pi(C)}{m}} \cdot \frac{(S_{m\tau_C^m(p_n)+1} - S_m)}{tr_n}\right.\right.\right.$$

$$\left.\left.\left. - x_0\right\| < r,\ \tau_C^m(p_n) < \frac{1}{q_n}\left[\frac{n}{m}\right]\right\}\right) - b\rho P(A_1),$$

where the last step follows from (5.7). Define

$$\begin{cases} \sigma_C^m(0) = \inf\{n \geq 0;\ X_{mn} \in C\} \\ \sigma_C^m(k+1) = \inf\{n > \sigma_C^m(k);\ X_{mn} \in C\} \quad (k \geq 0). \end{cases}$$

Then, one can verify that

$$\left(S_{m\tau_C^m(p_n)+1} - S_m,\ \tau_C^m(p_n)\right) = \left(S_{m\sigma_C^m(p_n-1)+1},\ 1 + \sigma_C^m(p_n-1)\right) \circ \theta^m.$$

For any $x \in C$, therefore,

$$P_x\Big\{\Big\|\sqrt{\frac{\pi(C)}{m}} \cdot \frac{(S_{m\tau_C^m(p_n)+1} - S_m)}{tr_n} - x_o\Big\| < r, \ \tau_C^m(p_n) < \frac{1}{q_n}\Big[\frac{n}{m}\Big]\Big\}$$

$$= \int P^m(x, dy) P_y\Big\{\Big\|\sqrt{\frac{\pi(C)}{m}} \cdot \frac{S_{m\sigma_C^m(p_n-1)+1}}{tr_n} - x_o\Big\| < r,$$

$$\sigma_C^m(p_n - 1) < \frac{1}{q_n}\Big[\frac{n}{m}\Big] - 1\Big\}$$

$$\geq b P_\nu\Big\{\Big\|\sqrt{\frac{\pi(C)}{m}} \cdot \frac{S_{m\sigma_C^m(p_n-1)+1}}{tr_n} - x_o\Big\| < r, \ \sigma_C^m(p_n - 1) < \frac{1}{q_n}\Big[\frac{n}{m}\Big] - 1\Big\},$$

where the last step follows from (5.4). Note that $\nu(C) = 1$. Under the law P_ν, $\sigma_C^m(0) = 0$ a.s. and therefore,

$$\sigma_C^m(k) = \tau_C^m(k) \quad \text{a.s.} \ P_\nu \quad k = 1, 2, \cdots.$$

Hence, we have proved that

$$P_{X_{m\tau_C^m((q_n-1)p_n)}}\Big\{\Big\|\sqrt{\frac{\pi(C)}{m}} \cdot \frac{(S_{m\tau_C^m(p_n)+1} - S_m)}{tr_n} - x_o\Big\| < r,$$

$$\tau_C^m(p_n) < \frac{1}{q_n}\Big[\frac{n}{m}\Big]\Big\}$$

$$\geq b P_\nu\Big\{\Big\|\sqrt{\frac{\pi(C)}{m}} \cdot \frac{S_{m\tau_C^m(p_n-1)+1}}{tr_n} - x_o\Big\| < r,$$

$$\tau_C^m(p_n - 1) < \frac{1}{q_n}\Big[\frac{n}{m}\Big] - 1\Big\} \quad \text{a.s.}$$

By (5.21) we have

$$P_\mu\Big\{\Big\|\sqrt{\frac{\pi(C)}{m}} \cdot \frac{S_{m\tau_C^m(q_np_n)+1}}{tq_nr_n} - x_o\Big\| < 2r, \ \tau_C^m(q_np_n) < \Big[\frac{n}{m}\Big]\Big\}$$

$$\geq P_\mu\Big\{\Big\|\sqrt{\frac{\pi(C)}{m}} \cdot \frac{S_{m\tau_C^m((q_n-1)p_n)+1}}{t(q_n-1)r_n} - x_o\Big\| < 2r,$$

$$\tau_C^m((q_n-1)p_n) < \frac{q_n - 1}{q_n}\Big[\frac{n}{m}\Big]\Big\}$$

$$\times \Big(b P_\nu\Big\{\Big\|\sqrt{\frac{\pi(C)}{m}} \cdot \frac{S_{m\tau_C^m(p_n-1)+1}}{tr_n} - x_o\Big\| < r,$$

$$\tau_C^m(p_n - 1) < \frac{1}{q_n}\Big[\frac{n}{m}\Big] - 1\Big\} - \rho\Big).$$

In view of (5.19), repeating the above procedure we obtain

$$P_\mu\left\{\left\|\sqrt{\frac{\pi(C)}{m}} \cdot \frac{S_{m\tau_C^m(q_n p_n)+1}}{tq_n r_n} - x_o\right\| < 2r, \ \tau_C^m(q_n p_n) < \left[\frac{n}{m}\right]\right\}$$

$$\geq P_\mu\left\{\left\|\sqrt{\frac{\pi(C)}{m}} \cdot \frac{S_{m\tau_C^m(p_n)+1}}{tr_n} - x_o\right\| < 2r, \ \tau_C^m(p_n) < \frac{1}{q_n}\left[\frac{n}{m}\right]\right\}$$

$$\times \left(bP_\nu\left\{\left\|\sqrt{\frac{\pi(C)}{m}} \cdot \frac{S_{m\tau_C^m(p_n-1)+1}}{tr_n} - x_o\right\| < r, \right.\right.$$

$$\left.\left. \tau_C^m(p_n-1) < \frac{1}{q_n}\left[\frac{n}{m}\right] - 1\right\} - \rho\right)^{q_n - 1}$$

for sufficiently large n. By definition $q_n \sim t^{-2}\lambda^{-1}\pi(C)n^{-1}b_n^2$ as $n \to \infty$. Taking (5.8), (5.13) and (5.18) into account gives

$$\liminf_{n\to\infty} \frac{n}{b_n^2} \log P_\mu\left\{\left\|\sqrt{\frac{\pi(C)}{m}} \cdot \frac{S_{m\tau_C^m(q_n p_n)+1}}{tq_n r_n} - x_o\right\| < 2r, \right.$$

$$\left. \tau_C^m(q_n p_n) < \left[\frac{n}{m}\right]\right\}$$

$$\geq \frac{\pi(C)}{\lambda t^2} \log\left(b\gamma(tB_{x_o,r}) - \rho\right).$$

This, together with (5.20), implies that

$$\liminf_{n\to\infty} \frac{n}{b_n^2} \log P_\mu\left\{\frac{S_n}{b_n} \in G\right\} \geq \frac{\pi(C)}{\lambda t^2} \log\left(b\gamma(tB_{x_o,r}) - \rho\right).$$

Letting $\rho \longrightarrow 0^+$ and $\lambda \longrightarrow \pi(C)^-$ gives (5.5). ∎

Appendix

1. Proof of Theorem I-2.2

Let $\{\tilde{\mathcal{A}}_k\}_{k\geq 0}$ be the filtration given in (I-2.11) and define

$$\begin{cases} \tau_{\alpha^*}(1) = \inf\{n \geq 1;\ \Phi_n \in \alpha^*\}, \\ \tau_{\alpha^*}(k+1) = \inf\{n > \tau_\alpha^*(k);\ \Phi_n \in \alpha^*\} \quad (k \geq 1). \end{cases}$$

Let $\tilde{\theta}\colon \prod_{n=0}^{\infty}(E^m \times I) \longrightarrow \prod_{n=0}^{\infty}(E^m \times I)$ be the shift operator w.r.t. the split chain given in (I-2.7). Then for each $j \geq 0$ and $k \geq 1$,

$$\tau(j) + \tau_{\alpha^*}(k) \circ \tilde{\theta}^{\tau(j)} = \tau(j+k) \quad \text{a.s.}$$

Let

$$\Delta B'_k = (X_m, \cdots, X_{m\tau_{\alpha^*}(n_k - n_{k-1})}) \quad k = 1, 2, \cdots.$$

By (2.9), for each $k \geq 1$,

$$\Delta B'_k \circ \tilde{\theta}^{\tau(n_{k-1})} = \left(X_m \circ \tilde{\theta}^{\tau(n_{k-1})}, \cdots, X_{m\tau_{\alpha^*}(n_k - n_{k-1})} \circ \tilde{\theta}^{\tau(n_{k-1})}\right)$$
$$= \left(X_{m+m\tau(n_{k-1})}, \cdots, X_{m\tau_{\alpha^*}(n_k - n_{k-1}) \circ \tilde{\theta}^{\tau(n_{k-1})} + m\tau(n_{k-1})}\right)$$
$$= (X_{m(\tau(n_{k-1})+1)}, X_{m(\tau(n_{k-1})+1)+1}, \cdots, X_{m\tau(n_k)}) = B_k$$

Given $k \geq 1$, let $A \in \tilde{\mathcal{A}}_{\tau(n_{k-1})}$ and let C_k be a measurable set in the "space of blocks" (formulated in the obvious way). Using the strong Markov property ((I-2.12)) gives

$$\tilde{P}_{\mu^*}(A \cap \{B_k \in C_k\}) = \tilde{P}_{\mu^*}(A \cap \{\Delta B'_k \circ \tilde{\theta}^{\tau(n_{k-1})} \in C_k\})$$
$$= \tilde{E}_{\mu^*}\left(I_A \cdot I_{C_k}(\Delta B'_k \circ \tilde{\theta}^{\tau(n_{k-1})}) | \tilde{\mathcal{A}}_{\tau(n_{k-1})}\right)$$
$$= \tilde{E}_{\mu^*}\left(I_A \cdot \tilde{P}_{\Phi_{\tau(n_{k-1})}}\{\Delta B'_k \in C_k\}\right).$$

Inductively, one can prove that

$$\tau_{\alpha^*}(k) = 1 + \tau(k-1) \circ \tilde{\theta} \quad \text{a.s.} \quad (k \geq 1).$$

Consequently, by (I-2.9) we have

$$\Delta B'_k = \Delta B_k \circ \tilde{\theta} \quad \text{a.s.} \quad (k \geq 1),$$

where

$$\Delta B_k = (X_0, \cdots, X_{m\tau(n_k - n_{k-1} - 1)}).$$

In view of (I-2.21),

$$\tilde{P}_{\Phi_{\tau(n_{k-1})}}\{\Delta B'_k \in C_k\} = \tilde{P}_{\nu^*}\{\Delta B_k \in C_k\} \quad \text{a.s.}$$

Hence, we have proved that

$$\tilde{P}_{\mu^*}(A \cap \{B_k \in C_k\}) = \tilde{P}_{\mu^*}(A) \cdot \tilde{P}_{\nu^*}\{\Delta B_k \in C_k\}.$$

Iterating this procedure we have that for measurable sets $C_0, C_1, \cdots, C_k$ in the blocks space,

$$\tilde{P}_{\mu^*}\{B_0 \in C_0, \ B_1 \in C_1, \cdots, B_k \in C_k\}$$
$$= \tilde{P}_{\mu^*}\{B_0 \in C_0\} \cdot \tilde{P}_{\nu^*}\{\Delta B_1 \in C_1\} \cdots \tilde{P}_{\nu^*}\{\Delta B_k \in C_k\}.$$

Hence, the desired conclusion is valid. ∎

2. Proof of Theorem I-2.5

Let $h: E \times I \longrightarrow \mathbf{R}$ be a measurable function with $||h||_\infty \leq 1$ and define $h^*: E \longrightarrow \mathbf{R}$ by

$$h^*(x) = (1 - bI_C(x))h(x, 0) + bI_C(x)h(x, 1) \quad x \in E.$$

Clearly, $||h^*||_\infty \le 1$. In view of (I-2.1) and (I-2.14), by Lemma 3.1-(i) of de Acosta (1988) we have

$$\bar{P}^n h(x,y) = \int \bar{P}^n((x,y), d(x_1, y_1)) h(x_1, y_1)$$
$$= \int \bar{P}^n((x,y), d(x_1, y_1)) I_0(y_1) h(x_1, 0)$$
$$+ \int \bar{P}^n((x,y), d(x_1, y_1)) I_1(y_1) h(x_1, 1)$$
$$= I_0(y) \int Q(x, dx_1) \int P^{m(n-1)}(x_1, dx_2)(1 - bI_C(x_2)) h(x_2, 0)$$
$$(A.1) \quad + I_1(y) \int \nu(dx_1) \int P^{m(n-1)}(x_1, dx_2)(1 - bI_C(x_2)) h(x_2, 0)$$
$$+ I_0(y) \int Q(x, dx_1) \int P^{m(n-1)}(x_1, dx_2) bI_C(x_2) h(x_2, 1)$$
$$+ I_1(y) \int \nu(dx_1) \int P^{m(n-1)}(x_1, dx_2) bI_C(x_2) h(x_2, 1)$$
$$= I_0(y) \int Q(x, dx_1) P^{m(n-1)} h^*(x_1)$$
$$+ I_1(y) \int \nu(dx_1) P^{m(n-1)} h^*(x_1).$$

By definition one can verify that

$$(A.2) \quad \pi^*(h) \equiv \int h(x,y) \pi^*(d(x,y)) = \int h^*(x) \pi(dx) \equiv \pi(h^*).$$

Therefore,

$$|\bar{P}^n h(x,y) - \pi^*(h)|$$
$$\le I_0(y) \int Q(x, dx_1) ||P^{m(n-1)}(x_1, \cdot) - \pi||_V$$
$$+ I_1(y) \int \nu(dx_1) ||P^{m(n-1)}(x_1, \cdot) - \pi||_V.$$

Since h is arbitrary,

$$||\bar{P}^n((x,y), \cdot) - \pi^*||_V$$
$$(A.3) \quad \le I_0(y) \int Q(x, dx_1) ||P^{m(n-1)}(x_1, \cdot) - \pi||_V$$
$$+ I_1(y) \int \nu(dx_1) ||P^{m(n-1)}(x_1, \cdot) - \pi||_V \quad \forall (x,y) \in E \times I.$$

Hence, (I-1.6) implies that

$$\lim_{n\to\infty} ||\bar{P}^n((x,y),\cdot) - \pi^*||_V = 0 \quad \forall (x,y) \in E \times I.$$

Thus (i) is proved.

In view of (I-2.1) and (I-2.3), integrating both sides of (A.1) yields:

$$
\begin{aligned}
\mu^* \bar{P}^n h &= \int \mu(dx)(1 - bI_C(x)) \int Q(x, dx_1) P^{m(n-1)} h^*(x_1) \\
&\quad + \int \mu(dx) bI_C(x) \int \nu(dx_1) P^{m(n-1)} h^*(x_1) \\
&= \int \mu(dx) \int P^m(x, dx_1) P^{m(n-1)} h^*(x_1) = \mu P^{mn} h^*.
\end{aligned}
$$
(A.4)

This, along with (A.2), implies that

$$|\mu^* \bar{P}^n h - \pi^*(h)| = |\mu P^{mn} h^* - \pi(h^*)| \le ||\mu P^{mn} - \pi||_V.$$

Hence,

(A.5) $$||\mu^* \bar{P}^n - \pi^*||_V \le ||\mu P^{mn} - \pi||_V.$$

On the other hand, if we take $h\colon E \longrightarrow \mathbf{R}$ depending only on $x \in E$, then $h^* = h$. Applying (A.2) and (A.4) again we have that

$$|\mu P^{mn} h - \pi(h)| \le ||\mu^* \bar{P}^n - \pi^*||_V.$$

Consequently,

(A.6) $$||\mu P^{mn} - \pi||_V \le ||\mu^* \bar{P}^n - \pi^*||_V.$$

Therefore, (ii) follows from (A.5) and (A.6).

From (I-2.1) and (A.3) we have

$$\int \pi^*(d(x,y))||\bar{P}^n((x,y),\cdot) - \pi^*||_V$$

$$\leq \int \pi(dx)(1 - bI_C(x))\int Q(x,dx_1)||P^{m(n-1)}(x_1,\cdot) - \pi||_V$$

(A.7) $$+ \int \pi(dx)bI_C(x)\int \nu(dx_1)||P^{m(n-1)}(x_1,\cdot) - \pi||_V$$

$$= \int \pi(dx) \int P^m(x,dx_1)||P^{m(n-1)}(x_1,\cdot) - \pi||_V$$

$$= \int \pi(dx)||P^{m(n-1)}(x,\cdot) - \pi||_V.$$

We now take $h: E \longrightarrow \mathbf{R}$ depending only on $x \in E$ and take $\mu = \delta_x$ in (A.4). Then $h^* = h$ and

$$P^{mn}h(x) = (1 - bI_C(x))\bar{P}^n h(x,0) + bI_C(x)\bar{P}^n h(x,1).$$

This, together with (A.2), implies that

$$|P^{mn}h(x) - \pi(h)|$$

$$\leq (1 - bI_C(x))||\bar{P}^n((x,0),\cdot) - \pi^*||_V$$

$$+ bI_C(x)||\bar{P}^n((x,1),\cdot) - \pi^*||_V.$$

Hence

$$||P^{mn}(x,\cdot) - \pi||_V$$

$$\leq (1 - bI_C(x))||\bar{P}^n((x,0),\cdot) - \pi^*||_V$$

$$+ bI_C(x)||\bar{P}^n((x,1),\cdot) - \pi^*||_V$$

for all $x \in E$. Consequently,

$$\int \pi(dx)||P^{mn}(x,\cdot) - \pi||_V$$

(A.8) $$\leq \int \pi(dx)(1 - bI_C(x))||\bar{P}^n((x,0),\cdot) - \pi^*||_V$$

$$+ \int \pi(dx)bI_C(x)||\bar{P}^n((x,1),\cdot) - \pi^*||_V$$

$$= \int \pi^*(d(x,y))||\bar{P}^n((x,y),\cdot) - \pi^*||_V.$$

Therefore, (iii) follows from (A.7) and (A.8). ∎

3. Proof of Theorem I-3.4

We only prove (I-3.11) in the case $m > 1$, as the proof of the rest is analogous. Define

$$T = \inf\{k \geq 0;\ \|S_{m\tau(k)+1} - x_0\| < t\}.$$

Then from (I-2.8),

$$P_\mu\{\|S_n - x_0\| < s+t\} = \tilde{P}_{\mu^*}\{\|S_n - x_0\| < s+t\}$$
$$\geq \tilde{P}_{\mu^*}\{\|S_n - x_0\| < s+t\ \ T < i(n)\}$$
$$= \sum_{k=0}^{\infty} \tilde{P}_{\mu^*}\{\|S_n - x_0\| < s+t\ \ T = k,\ i(n) \geq k+1\}$$
$$\geq \sum_{k=0}^{\infty} \tilde{P}_{\mu^*}\{\|S_n - S_{m\tau(k)+1}\| < s\ \ T = k,\ i(n) \geq k+1\}.$$

By definition,

$$\{i(n) \geq k+1\} = \left\{\tau(k) \leq \left[\frac{n}{m}\right] - 1\right\} \quad (k \geq 0).$$

Therefore,

$$P_\mu\{\|S_n - x_0\| < s+t\} = \tilde{P}_{\mu^*}\{\|S_n - x_0\| < s+t\}$$
$$\geq \sum_{k=0}^{\infty} \tilde{P}_{\mu^*}\left\{\|S_n - S_{m\tau(k)+1}\| < s\ \ T = k,\ \tau(k) \leq \left[\frac{n}{m}\right] - 1\right\}$$
$$= \sum_{k=0}^{\infty} \sum_{j=0}^{[n/m]-1} \tilde{P}_{\mu^*}\{T = k,\ \tau(k) = j,\ \|S_n - S_{mj+1}\| < s\}$$

(A.9)
$$= \sum_{k=0}^{\infty} \sum_{j=0}^{[n/m]-1} \tilde{E}_{\mu^*}\left(I_{\{T=k,\ \tau(k)=j\}} \cdot \tilde{P}_{\Phi_j}\left\{\|\sum_{i=1}^{n-mj} \xi(X_j)\| < s\right\}\right)$$
$$\geq \sum_{k=0}^{\infty} \sum_{j=0}^{[n/m]-1} \tilde{P}_{\mu^*}\{T = k,\ \tau(k) = j\}$$
$$\times \inf_{x \in C} \tilde{P}_{(x,1)}\left\{\|\sum_{i=1}^{n-mj} \xi(X_j)\| < s\right\},$$

where the second equality follows from the Markov property ((I-2.10)). By (I-1.4) and (I-2.4), for all $x \in C$ we have

$$\tilde{P}_{(x,1)}\left\{\sum_{i=1}^{m-1}\|\xi(X_i)\| > \frac{s}{2}\right\}$$
$$= P_1\left(x, \left\{\sum_{i=1}^{m-1}\|\xi(x_i)\| > \frac{s}{2}\right\}\right) \leq \frac{1}{b}P_x\left\{\sum_{i=1}^{m-1}\|\xi(X_i)\| > \frac{s}{2}\right\}.$$

Hence for j satisfying $n - mj > m - 1$ and for $x \in C$,

$$(A.10)\quad\begin{aligned}&\tilde{P}_{(x,1)}\left\{\|\sum_{i=1}^{n-mj}\xi(X_j)\| < s\right\}\\ &\geq \tilde{P}_{(x,1)}\left\{\|\sum_{i=m}^{n-mj}\xi(X_j)\| < \frac{s}{2}\right\} - \tilde{P}_{(x,1)}\left\{\sum_{i=1}^{m-1}\|\xi(X_i)\| > \frac{s}{2}\right\}\\ &\geq \tilde{E}_{(x,1)}\left(\tilde{P}_{\Phi_1}\left\{\|S_{n-mj-m-1}\| < \frac{s}{2}\right\}\right) - \frac{1}{b}P_x\left\{\sum_{i=1}^{m-1}\|\xi(X_i)\| > \frac{s}{2}\right\}\\ &= P_\nu\left\{\|S_{n-mj-m-1}\| < \frac{s}{2}\right\} - \frac{1}{b}P_x\left\{\sum_{i=1}^{m-1}\|\xi(X_i)\| > \frac{s}{2}\right\} \geq r,\end{aligned}$$

where the equality follows from (I-2.8) (with $\mu = \nu$) and (I-2.17).

For j satisfying $n - mj \leq m - 1$ and for $x \in C$,

$$(A.11)\quad \tilde{P}_{(x,1)}\left\{\|\sum_{i=1}^{n-mj}\xi(X_j)\| < s\right\} \geq 1 - \tilde{P}_{(x,1)}\left\{\sum_{i=1}^{m-1}\|\xi(X_i)\| > s\right\} \geq r.$$

From (A.9), (A.10) and (A.11),

$$P_\mu\{\|S_n - x_0\| < s + t\} \geq r\sum_{k=0}^{\infty}\sum_{j=0}^{[n/m]-1}\tilde{P}_{\mu^*}\{T = k, \tau(k) = j\}$$
$$= r\tilde{P}_{\mu^*}\left\{\min_{0 \leq k < i(n)}\|S_{m\tau(k)+1} - x_0\| < t\right\}.$$

∎

4. Proof of Theorem I-4.1

The proof of Theorem 4.1 consists of several lemmas. Our first lemma was initially used by Burkholder-Gundy (1970) and later used extenssively in many different situations. For completeness we include it here.

Lemma A.1. *Let U and V be non-negative random variables. Suppose there exist positive reals β, δ, γ and t_o such that $\beta^{-1} - \gamma > 0$ and*

$$(A.12) \qquad P\{U \geq \beta t, \ V < \delta t\} \leq \gamma P\{U \geq t\} \quad \text{for all } t \geq t_o.$$

Then,

$$(A.13) \qquad EU \leq (\beta^{-1} - \gamma)^{-1}(t_o + \delta^{-1} EV).$$

Proof. For each $t \geq t_o$,

$$P\{U \geq \beta t\} \leq P\{U \geq \beta t, \ V < \delta t\} + P\{V \geq \delta t\}$$
$$\leq \gamma P\{U \geq t\} + P\{V \geq \delta t\},$$

Integrating both sides gives

$$EU = \beta \int_0^{+\infty} P\{U \geq \beta t\} dt \leq \beta \left[t_o + \int_{t_o}^{+\infty} P\{U \geq \beta t\} dt \right]$$
$$\leq \beta \left[t_o + \gamma \int_{t_o}^{+\infty} P\{U \geq t\} dt + \int_{t_o}^{+\infty} P\{V \geq \delta t\} dt \right]$$
$$\leq \beta \left[t_o + \gamma EU + \delta^{-1} EV \right].$$

Solving for EU gives (A.13). ∎

We may assume that φ is a non-decreasing continuous function such that for some fixed $\lambda > 0$,

$$(A.14) \qquad \varphi(cs) \leq c^\lambda \varphi(s) \quad \text{for all } s \geq 0 \text{ and } c \geq 1$$

for, there is a non-decreasing continuous function $\hat{\varphi}\colon [0,+\infty) \longrightarrow [0,+\infty)$ (Lemma A.2 in [29]) satisfying (A.14) for some $\hat{\lambda} > 0$ such that

$$\hat{\varphi}(s) \leq \varphi(s) \leq 2^{\lambda+1}\hat{\varphi}(s) \quad \forall s \geq 0,$$

so we can take $\hat{\varphi}$ instead of φ if otherwise. So (A.14) is assumed in the rest of this section.

Lemma A.2. *Let $\{X_n\}_{n\geq 0}$ be a Harris recurrent Markov chain. Assume that the set $C \in \mathcal{E}$ can be written as a union of finitely many small sets and*

$$(A.15) \qquad \int_C \pi(dx) E_x \max_{n \leq \tau_C} \varphi(||S_n||) < +\infty.$$

Then, for every $D \in \mathcal{S}$ with $D \subset C$, we have $D \in \mathcal{S}_\varphi(\xi)$, i.e.,

$$(A.16) \qquad \int_D \pi(dx) E_x \max_{n \leq \tau_D} \varphi(||S_n||) < +\infty.$$

Proof. By assumption there exist small sets $C_1, \cdots, C_N$ such that

$$(A.17) \qquad C = \bigcup_{i=1}^{N} C_i.$$

By Theorem 5.2.2 in Meyn-Tweedie (1993), for each $1 \leq i \leq N$ there exist $b_i > 0$, $m_i \geq 1$ and probability measure ν_i on $(E, \mathcal{E})$ with $\nu_i(C_i) > 0$ such that $P^{m_i} \geq b_i I_{C_i} \otimes \nu_i$. Consider the subset I_i of positive integers given by

$$I_i = \{m \geq 1;\ P^m \geq b I_{C_i} \otimes \nu_i \text{ for some } b > 0\}.$$

As we mention in Section 1, the greatest common divisor of I_i is the period d of the chain $\{X_n\}_{n\geq 0}$, which is of course independent of i. Besides, I_i is closed under addition: Let $n_1 \geq 1$ and $n_2 \geq 1$ be such that

$$P^{n_1} \geq \beta_1 I_{C_i} \otimes \nu_i \text{ and } P^{n_2} \geq \beta_2 I_{C_i} \otimes \nu_i.$$

By the Chapman-Kolmogorov equation,

$$P^{n_1+n_2}(x,\cdot) = \int P^{n_1}(x,dx_1)P^{n_2}(x_1,\cdot) \geq \int_{C_i} P^{n_1}(x,dx_1)P^{n_2}(x_1,\cdot)$$
$$\geq P^{n_1}(x,C_i)\beta_2\nu_i(\cdot) \geq (\beta_1\beta_2\nu_i(C_i))I_{C_i}\otimes\nu_i(\cdot).$$

According to p.527, Lemma D.7.4 in Meyn-Tweedie (1993), $nd \in I_i$ for all sufficiently large integer n. In particular, there exist $b > 0$ and $m \geq 1$ such that

(A.18) $$P^m \geq bI_{C_i} \otimes \nu_i \quad i = 1,\cdots,N.$$

Define the regeneration times $\{\tau_C(k)\}_{k\geq 0}$ as follows

(A.19) $$\begin{cases} \tau_C(0) = 0 \text{ and } \tau_C(1) = \tau_C \\ \tau_C(k+1) = \inf\{n > \tau_C(k);\ X_n \in C\} \quad (k \geq 1). \end{cases}$$

Define

$$T = \inf\{k \geq 1;\ X_{\tau_C(k)} \in D\}.$$

Then,

(A.20) $$\tau_C(T) = \tau_D \quad \text{a.s.}$$

Choose $0 < \epsilon < 1/2$ such that

(A.21) $$\left(1 - \frac{b}{2}\right)/(1-2\epsilon)^\lambda < 1$$

and let $s \geq 0$ be fixed but arbitrary. Define

$$\eta_k = \max_{\tau_C(k-1)\leq n\leq m+\tau_C(k)} \|S_n - S_{\tau_C(k-1)}\| \quad k = 1,2,\cdots,$$

$$\rho = \inf\{k \geq 1;\ \max_{n\leq \tau_C(k)} \|S_n\| > (1-2\epsilon)s\}.$$

According to Proposition 5.6-(ii) in Nummelin (1984), $0 < \pi(D) < +\infty$. Let π_D (see (I-1.5) for a definition of π_D) be the initial distribution. In view of (A.20) we have

$$P_{\pi_D}\{\max_{n\leq \tau_D} \|S_n\| \geq s, \max_{1\leq k\leq T} \eta_k \leq \epsilon s\}$$
$$= P_{\pi_D}\{\max_{n\leq \tau_D} \|S_n\| \geq s, \max_{1\leq k\leq T} \eta_k \leq \epsilon s, \ \rho \leq T\}$$
$$= \sum_{l=1}^{\infty} P_{\pi_D}\{\rho = l, \ T \geq l, \ \max_{n\leq \tau_D} \|S_n\| \geq s, \max_{1\leq k\leq T} \eta_k \leq \epsilon s\}.$$

Write

$$\tau'_D = \inf\{n \geq m+1; \ X_n \in D\}.$$

One can easily see that for each $l \geq 1$,

$$\tau_D \leq \tau_C(l) + \tau'_D \circ \theta^{\tau_C(l)} \quad \text{a.s.}$$

Hence,

$$P_{\pi_D}\{\max_{n\leq \tau_D} \|S_n\| \geq s, \max_{1\leq k\leq T} \eta_k \leq \epsilon s\}$$
$$\leq \sum_{l=1}^{\infty} P_{\pi_D}\{\rho = l, \ T \geq l, \max_{n\leq \tau_C(l)+\tau'_D \circ \theta^{\tau_C(l)}} \|S_n\| \geq s, \max_{1\leq k\leq T} \eta_k \leq \epsilon s\}$$
$$\leq \sum_{l=1}^{\infty} P_{\pi_D}\{\rho = l, \ T \geq l,$$
$$\max_{m+\tau_C(l)\leq n\leq \tau_C(l)+\tau'_D \circ \theta^{\tau_C(l)}} \|S_n - S_{m+\tau_C(l)}\| \geq \epsilon s\},$$

where the last step follows from the fact that on the event $\{\rho = l, \ T \geq l\}$, we have for all $m + \tau_C(l) \leq n \leq \tau_C(l) + \tau'_D \circ \theta^{\tau_C(l)}$ that

$$\|S_n\| \leq \|S_n - S_{m+\tau_C(l)}\| + \max_{k\leq \tau_C(l-1)} \|S_k\| + \eta_l$$
$$\leq \|S_n - S_{m+\tau_C(l)}\| + (1-2\epsilon)s + \max_{1\leq k\leq T} \eta_k.$$

Note that for each $l \geq 1$,

$$\max_{m+\tau_C(l) \leq n \leq \tau_C(l)+\tau'_D \circ \theta^{\tau_C(l)}} \|S_n - S_{m+\tau_C(l)}\|$$
$$= \left(\max_{m \leq n \leq \tau'_D} \|S_n - S_m\|\right) \circ \theta^{\tau_C(l)} \quad \text{a.s.}$$

By the strong Markov property we obtain

$$P_{\pi_D}\{\max_{n \leq \tau_D} \|S_n\| \geq s, \max_{1 \leq k \leq T} \eta_k \leq \epsilon s\}$$
$$\leq \sum_{l=1}^{\infty} E_{\pi_D}\left(I_{\{\rho=l,\ T \geq l\}} \cdot P_{X_{\tau_C(l)}}\{\max_{m \leq n \leq \tau'_D} \|S_n - S_m\| \geq \epsilon s\}\right).$$

For each $x \in C$,

$$P_x\{\max_{m \leq n \leq \tau'_D} \|S_n - S_m\| < \epsilon s\}$$
$$= \int P^m(x, dx_1) P_{x_1}\{\max_{n \leq \tau_D} \|S_n\| < \epsilon s\}$$
$$\geq b \min_{1 \leq i \leq N} P_{\nu_i}\{\max_{n \leq \tau_D} \|S_n\| < \epsilon s\},$$

where the first step follows from the Markov property and the second from (A.18). Consequently,

$$P_{X_{\tau_C(l)}}\{\max_{m \leq n \leq \tau'_D} \|S_n - S_m\| \geq \epsilon s\}$$
$$\leq 1 - b \min_{1 \leq i \leq N} P_{\nu_i}\{\max_{n \leq \tau_D} \|S_n\| < \epsilon s\} \quad \text{a.s.}$$

Therefore,

(A.22)
$$P_{\pi_D}\{\max_{n \leq \tau_D} \|S_n\| \geq s, \max_{1 \leq k \leq T} \eta_k \leq \epsilon s\}$$
$$\leq \left(1 - b \min_{1 \leq i \leq N} P_{\nu_i}\{\max_{n \leq \tau_D} \|S_n\| < \epsilon s\}\right) P_{\pi_D}\{\rho \leq T\}$$
$$= \left(1 - b \min_{1 \leq i \leq N} P_{\nu_i}\{\max_{n \leq \tau_D} \|S_n\| < \epsilon s\}\right)$$
$$\times P_{\pi_D}\{\max_{n \leq \tau_D} \|S_n\| > (1 - 2\epsilon)s\}.$$

Define the inverse function φ^{-1} of φ by

(A.23) $$\varphi^{-1}(t) = \inf\{s \geq 0;\ \varphi(s) \geq t\} \quad t \geq 0.$$

It is easy to see that

(A.24) $\quad \{s;\ s > \varphi^{-1}(t)\} \subset \{s;\ \varphi(s) \geq t\} \subset \{s;\ s \geq \varphi^{-1}(t)\} \quad \forall t \geq 0.$

By (A.14),

(A.25) $\quad \varphi\big((1-2\epsilon)^{-1}s\big) \leq (1-2\epsilon)^{-\lambda}\varphi(s) \quad \forall s \geq 0.$

Let $\beta = (1-2\epsilon)^{-\lambda}$. For given $t \geq 0$, bringing $s = \varphi^{-1}(\beta t)$ into (A.22) gives

$$P_{\pi_D}\Big\{ \max_{n \leq \tau_D} \varphi(\|S_n\|) \geq \beta t,\ \max_{1 \leq k \leq T} \varphi(\epsilon^{-1}\eta_k) < \beta t \Big\}$$
$$\leq \Big(1 - b \min_{1 \leq i \leq N} P_{\nu_i}\big\{ \max_{n \leq \tau_D} \|S_n\| < \epsilon \varphi^{-1}(\beta t) \big\}\Big)$$
$$\times P_{\pi_D}\Big\{ \max_{n \leq \tau_D} \varphi(\|S_n\|) \geq t \Big\} \quad \forall t \geq 0.$$

Take $t_o > 0$ sufficiently large that

$$\min_{1 \leq i \leq N} P_{\nu_i}\Big\{ \max_{n \leq \tau_D} \|S_n\| < \epsilon \varphi^{-1}(\beta t) \Big\} > \frac{1}{2} \quad \forall t \geq t_o.$$

In view of (A.21), Lemma A.1 applies if we take $\delta = \beta$, $\gamma = 1 - b/2$, $U = \max_{n \leq \tau_D} \varphi(\|S_n\|)$ and $V = \max_{1 \leq k \leq T} \varphi(\epsilon^{-1}\eta_k)$. Therefore, it remains to prove

$$E_{\pi_D} \max_{1 \leq k \leq T} \varphi(\epsilon^{-1}\eta_k) < +\infty.$$

By (A.14), it is enough to prove

(A.26) $\quad \displaystyle\int_D \pi(dx) E_x \sum_{k=1}^T \varphi(\eta_k) < +\infty.$

To do this, first we claim that

(A.27) $\quad \displaystyle\int_C \pi(dx) E_x \max_{n \leq m + \tau_C} \varphi(\|S_n\|) < +\infty.$

Note that $m + \tau_C \leq \tau_C(m+1)$ a.s. By the triangle inequality,

$$\max_{n \leq m+\tau_C} \|S_n\| \leq \max_{n \leq \tau_C(m+1)} \|S_n\| \leq \sum_{k=0}^{m} \max_{\tau_C(k) \leq n \leq \tau_C(k+1)} \|S_n - S_{\tau_C(k)}\| \quad \text{a.s.}$$

In view of (A.14), for any $s_0, \cdots, s_m \geq 0$,

$$\varphi(s_0 + \cdots + s_m) \leq \varphi\big((m+1) \max_{0 \leq k \leq m} s_k\big)$$
$$\leq (m+1)^\lambda \max_{0 \leq k \leq m} \varphi(s_k) \leq (m+1)^\lambda \big(\varphi(s_0) + \cdots + \varphi(s_m)\big).$$

Therefore,

$$\max_{n \leq m+\tau_C} \varphi(\|S_n\|) \leq (m+1)^\lambda \sum_{k=0}^{m} \max_{\tau_C(k) \leq n \leq \tau_C(k+1)} \varphi(\|S_n - S_{\tau_C(k)}\|) \quad \text{a.s.}$$

Hence,

$$\int_C \pi(dx) E_x \max_{n \leq m + \tau_C} \varphi(\|S_n\|)$$
$$\leq (m+1)^\lambda \sum_{k=0}^{m} \int_C \pi(dx) E_x \Big(\max_{\tau_C(k) \leq n \leq \tau_C(k+1)} \varphi(\|S_n - S_{\tau_C(k)}\|) \Big)$$
$$\leq (m+1)^\lambda \sum_{k=0}^{m} \int_C \pi(dx) E_x \Big(E_{X_{\tau_C(k)}} \max_{n \leq \tau_C} \varphi(\|S_n\|) \Big),$$

where the last step follows from the Markov property. Note that $\{X_{\tau_C(k)}\}_{k \geq 0}$ is a Harris recurrent Markov chain with state space C, transition probability P_C given by

$$P_C(x, A) = P_x\{X_{\tau_C} \in A\} \quad x \in C, \ A \subset C$$

and invariant measure $\pi(C \cap \cdot)$. In particular, for each $0 \leq k \leq m$,

$$\int_C \pi(dx) E_x \Big(E_{X_{\tau_C(k)}} \max_{n \leq \tau_C} \varphi(\|S_n\|) \Big) = \int_C \pi(dx) E_x \max_{n \leq \tau_C} \varphi(\|S_n\|) < +\infty.$$

Hence (A.27) holds.

We now come to the proof of (A.26). By the Markov property,

$$\int_D \pi(dx) E_x \sum_{k=1}^T \varphi(\eta_k) = \sum_{k=1}^\infty \int_D \pi(dx) E_x \varphi(\eta_k) I_{\{T \geq k\}}$$

$$= \sum_{k=1}^\infty \int_D \pi(dx) E_x \Big(I_{\{T \geq k\}} E_{X_{\tau_C(k-1)}} \max_{n \leq m+\tau_C} \varphi(\|S_n\|) \Big)$$

$$= \int_D \pi(dx) E_x \sum_{k=1}^T E_{X_{\tau_C(k-1)}} \Big(\max_{n \leq m+\tau_C} \varphi(\|S_n\|) \Big)$$

$$= \int_D \pi(dx) E_x \sum_{k=0}^{T-1} E_{X_{\tau_C(k)}} \Big(\max_{n \leq m+\tau_C} \varphi(\|S_n\|) \Big).$$

Applying Theorem 10.4.9 in Meyn-Tweedie (1993) to the chain $\{X_{\tau_C(k)}\}_{k \geq 0}$ gives

$$\int_D \pi(dx) E_x \sum_{k=0}^{T-1} E_{X_{\tau_C(k)}} \Big(\max_{n \leq m+\tau_C} \varphi(\|S_n\|) \Big)$$

$$= \int_C \pi(dx) E_x \max_{n \leq m+\tau_C} \varphi(\|S_n\|).$$

Therefore, the desired conclusion follows from (A.27). ∎

For $C \in \mathcal{E}^+$ and $m \geq 1$, write

(A.28) $$\tau_C^m = \inf\{n \geq 1;\ X_{nm} \in C\}.$$

From definition one can see that at time $m\tau_C^m$, C is hit by $\{X_{nm}\}_{n \geq 1}$. In particular, $\tau_C \leq m\tau_C^m$ a.s. However, we have the following:

Lemma A.3. Let $\{X_n\}_{n \geq 0}$ be an aperiodic Harris recurrent Markov chain and suppose $C \in \mathcal{S}_\varphi(\xi)$. Then, for any $m \geq 1$,

(A.29) $$\int_C \pi(dx) E_x \max_{n \leq m\tau_C^m} \varphi(\|S_n\|) < +\infty.$$

Proof. Consider the modular group Z_m:

$$Z_m = \{\hat{0}, \hat{1}, \cdots, \widehat{m-1}\},$$

where we use $\hat{k}$ to denote the equivalent class:

$$\hat{k} = \{n;\ n = k\ (mod)\ m\}.$$

One can easily verify that $\{(\hat{n}, X_n)\}_{n \geq 0}$ is a Markov chain with state space $Z_m \times E$ and the transition probability $\hat{P}$ given by

$$\hat{P}((\hat{k}, x), \Delta \times A) = \delta_{\hat{k}+\hat{1}}(\Delta) \cdot P(x, A).$$

According to Theorem 9.1.6 in Meyn-Tweedie (1993), $\{X_{nm}\}_{n \geq 0}$ is a Harris recurrent Markov chain. One can see how this implies the Harris recurrence of $\{(\hat{n}, X_n)\}_{n \geq 0}$. One can also see that $\{(\hat{n}, X_n)\}_{n \geq 0}$ has the invariant measure $u \otimes \pi$, where u is the uniform distribution on Z_m:

$$u(\hat{k}) = \frac{1}{m} \quad k = 0, 1, \cdots, m-1.$$

By assumption,

$$\int_{Z_m \times C} u \otimes \pi \big(d(\hat{k}, x)\big) \hat{E}_{(\hat{k},x)} \max_{n \leq \tau_{Z_m \times C}} \varphi(||S_n||)$$
$$= \int_C \pi(dx) E_x \max_{n \leq \tau_C} \varphi(||S_n||) < +\infty.$$

Note that for each $0 \leq k \leq m-1$, $\{\hat{k}\} \times C$ is a small set of the chain $\{(\hat{n}, X_n)\}_{n \geq 0}$ and that

$$Z_m \times C = \bigcup_{k=0}^{m-1} \{\hat{k}\} \times C.$$

Applying Lemma A.2 to the chain $\{(\hat{n}, X_n)\}_{n\geq 0}$ gives

$$\int_{\{\hat{0}\}\times C} u \otimes \pi\big(d(\hat{k}, x)\big) \hat{E}_{(\hat{k},x)} \max_{n\leq \tau_{\{\hat{0}\}\times C}} \varphi(\|S_n\|) < +\infty,$$

which is equivalent to (A.29). ∎

Let us mention the fact (Corollary 2.1-(iii) of Nummelin (1984)) that $\mathcal{S}$ is closed under finite unions when $\{X_n\}_{n\geq 0}$ is aperiodic. Therefore, the following lemma can be viewed as a converse of Lemma A.2:

Lemma A.4. *Let $\{X_n\}_{n\geq 0}$ be an aperiodic Harris recurrent Markov chain. Then for any $C, D \in \mathcal{S}$ with $D \subset C$, $D \in \mathcal{S}_\varphi(\xi)$ implies $C \in \mathcal{S}_\varphi(\xi)$.*

Proof. By assumption there exist $m \geq 1$, $b > 0$ and probability measure ν on $(E, \mathcal{E})$ such that

$$(A.30) \qquad P^m \geq b I_C \otimes \nu.$$

It is enough to prove

$$(A.31) \qquad \int_C \pi(dx) E_x \max_{n \leq m\tau_C^m} \varphi(\|S_n\|) < +\infty.$$

Let

$$(A.32) \qquad \begin{cases} \tau_C^m(0) = 0 \text{ and } \tau_C^m(1) = \tau_C^m \\ \tau_C^m(k+1) = \inf\{n > \tau_C^m(k);\ X_{mn} \in C\} \quad (k \geq 1). \end{cases}$$

Define

$$\varphi_k = \max_{m\tau_C^m(k-1) \leq n \leq m\tau_C^m(k)} \varphi(\|S_n - S_{m\tau_C^m(k-1)}\|) \quad k \geq 1,$$

$$T = \inf\{k \geq 1;\ X_{m\tau_C^m(k)} \in D\}.$$

Then,

(A.33) $$\tau_C^m(T) = \tau_D^m \quad \text{a.s.}$$

Let $t \geq 0$ be fixed but arbitrary and define

$$\rho = \inf\{k \geq 1; \ \sum_{j=1}^{k} \varphi_j I_{\{T \geq j\}} > t\}.$$

By (A.14) and (A.33), for each $k \geq 1$,

(A.34) $$\varphi_k I_{\{T \geq k\}} \leq \max_{n \leq m\tau_C^m(T)} \varphi(2\|S_n\|) \leq 2^\lambda \max_{n \leq m\tau_D^m} \varphi(\|S_n\|).$$

Choose $\epsilon > 0$ such that

(A.35) $$(1 - \frac{b}{2})(1 + 2\epsilon) < 1.$$

By the definition of ρ,

$$\{\rho < +\infty\} = \{\sum_{j=1}^{\infty} \varphi_j I_{\{T \geq j\}} > t\} \supset \{\sum_{j=1}^{\infty} \varphi_j I_{\{T \geq j\}} \geq (1 + 2\epsilon)t\}.$$

Therefore,

$$P_{\pi_D}\Big\{\sum_{j=1}^{\infty} \varphi_j I_{\{T \geq j\}} \geq (1+2\epsilon)t, \ \max_{n \leq m\tau_D^m} \varphi(\|S_n\|) \leq 2^{-\lambda-1}\epsilon t\Big\}$$

$$= P_{\pi_D}\Big\{\rho < +\infty, \ \sum_{j=1}^{\infty} \varphi_j I_{\{T \geq j\}} \geq (1+2\epsilon)t,$$

$$\max_{n \leq m\tau_D^m} \varphi(\|S_n\|) \leq 2^{-\lambda-1}\epsilon t\Big\}$$

$$= \sum_{k=1}^{\infty} P_{\pi_D}\Big\{\rho = k, \ \sum_{j=1}^{\infty} \varphi_j I_{\{T \geq j\}} \geq (1+2\epsilon)t,$$

$$\max_{n \leq m\tau_D^m} \varphi(\|S_n\|) \leq 2^{-\lambda-1}\epsilon t\Big\}$$

$$\leq \sum_{k=1}^{\infty} P_{\pi_D}\Big\{\rho = k, \ \sum_{j=k+2}^{\infty} \varphi_j I_{\{T \geq j\}} \geq \epsilon t\Big\},$$

where the last step follows from (A.34) and the fact that on the event $\{\rho = k\}$,

$$\sum_{j=1}^{\infty} \varphi_j I_{\{T \geq j\}}$$

$$\leq \sum_{j=1}^{k-1} \varphi_j I_{\{T \geq j\}} + \varphi_k I_{\{T \geq k\}} + \varphi_{k+1} I_{\{T \geq k+1\}} + \sum_{j=k+2}^{\infty} \varphi_j I_{\{T \geq j\}}$$

$$\leq t + 2^{1+\lambda} \max_{n \leq m\tau_D^m} \varphi(\|S_n\|) + \sum_{j=k+2}^{\infty} \varphi_j I_{\{T \geq j\}}.$$

For each $k \geq 0$, define

$$\hat{T}_k = \inf\{l \geq k+2; \ X_{m\tau_C^m(l)} \in D\}.$$

Since for each $k \geq 1$, $T \leq \hat{T}_k$ a.s., we have

$$P_{\pi_D}\left\{\sum_{j=1}^{\infty} \varphi_j I_{\{T \geq j\}} \geq (1+2\epsilon)t, \ \max_{n \leq m\tau_D^m} \varphi(\|S_n\|) \leq 2^{-\lambda-1}\epsilon t\right\}$$

(A.36) $\quad \leq \sum_{k=1}^{\infty} P_{\pi_D}\left\{\rho = k, \ \sum_{j=k+2}^{\infty} \varphi_j I_{\{\hat{T}_k \geq j\}} \geq \epsilon t\right\}$

$$= \sum_{k=1}^{\infty} P_{\pi_D}\left\{\rho = k, \ \sum_{j=k+2}^{\hat{T}_k} \varphi_j \geq \epsilon t\right\}.$$

Note that for each $j, k \geq 0$,

$$\tau_C^m(j) \circ \theta^{m\tau_C^m(k)} + m\tau_C^m(k) = \tau_C^m(j+k) \quad \text{a.s.}$$

By definition,

$$\varphi_j \circ \theta^{m\tau_C^m(k)} = \varphi_{j+k} \quad \text{a.s.}$$

and

$$k + \hat{T}_0 \circ \theta^{m\tau_C^m(k)} = k + \inf\{n \geq 2; \ X_{m\tau_C^m(n)} \circ \theta^{m\tau_C^m(k)} \in D\}$$

$$= k + \inf\{n \geq 2; \ X_{m\tau_C^m(n) \circ \theta^{m\tau_C^m(k)} + m\tau_C^m(k)} \in D\}$$

$$= k + \inf\{n \geq 2; \ X_{m\tau_C^m(n+k)} \in D\} = \hat{T}_k \quad \text{a.s.}$$

Consequently,

$$\left(\sum_{j=2}^{\hat{T}_0}\varphi_j\right)\circ\theta^{m\tau_C^m(k)} = \sum_{j=2}^{\hat{T}_0\circ\theta^{m\tau_C^m(k)}}\varphi_j\circ\theta^{m\tau_C^m(k)} = \sum_{j=2}^{\hat{T}_k-k}\varphi_{j+k} = \sum_{j=k+2}^{\hat{T}_k}\varphi_j \quad \text{a.s.}$$

Applying the Markov property to the right-hand side of (A.36) gives

(A.37)
$$P_{\pi_D}\left\{\sum_{j=1}^{\infty}\varphi_j I_{\{T\geq j\}} \geq (1+2\epsilon)t,\; \max_{n\leq m\tau_D^m}\varphi(\|S_n\|)\leq 2^{-\lambda-1}\epsilon t\right\}$$
$$\leq \sum_{k=1}^{\infty} E_{\pi_D}\left(I_{\{\rho=k\}}\cdot P_{X_{m\tau_C^m(k)}}\left\{\sum_{j=2}^{\hat{T}_0}\varphi_j\geq \epsilon t\right\}\right).$$

Define

$$\begin{cases} \sigma_C^m(0) = \inf\{n\geq 0;\; X_{mn}\in C\} \\ \sigma_C^m(k+1) = \inf\{n > \sigma_C^m(k);\; X_{mn}\in C\} \quad (k\geq 0), \end{cases}$$

$$\varphi_k' = \max_{m\sigma_C^m(k-1)\leq n\leq m\sigma_C^m(k)}\varphi(\|S_n - S_{m\sigma_C^m(k-1)}\|) \quad k\geq 1,$$

and

$$T' = \inf\{k\geq 1;\; X_{m\sigma_C^m(k)}\in D\}.$$

Inductively, one can prove that

$$\tau_C^m(k) = 1 + \sigma_C^m(k-1)\circ\theta^m \quad \text{a.s.} \quad (k\geq 1).$$

A similar computation gives

$$\left(\sum_{j=1}^{T'}\varphi_j'\right)\circ\theta^m = \sum_{j=2}^{\hat{T}_0}\varphi_j \quad \text{a.s.}$$

In view of (A.30), by the Markov property we have for every $x\in C$ that

$$P_x\left\{\sum_{j=2}^{\hat{T}_0}\varphi_j < \epsilon t\right\} = \int P^m(x,dx_1)P_{x_1}\left\{\sum_{j=1}^{T'}\varphi_j' < \epsilon t\right\} \geq bP_\nu\left\{\sum_{j=1}^{T'}\varphi_j' < \epsilon t\right\}.$$

Consequently,

$$P_{X_{m\tau_C^m(k)}}\Big\{\sum_{j=2}^{\hat{T}_0}\varphi_j \geq \epsilon t\Big\} \leq 1 - bP_\nu\Big\{\sum_{j=1}^{T'}\varphi'_j < \epsilon t\Big\} \quad \text{a.s.}$$

Bringing this into (A.37) gives

$$P_{\pi_D}\Big\{\sum_{j=1}^{\infty}\varphi_j I_{\{T\geq j\}} \geq (1+2\epsilon)t, \max_{n\leq m\tau_D^m}\varphi(\|S_n\|) \leq 2^{-\lambda-1}\epsilon t\Big\}$$

$$\leq \Big(1 - bP_\nu\Big\{\sum_{j=1}^{T'}\varphi'_j < \epsilon t\Big\}\Big)P_{\pi_D}\{\rho < +\infty\}$$

$$= \Big(1 - bP_\nu\Big\{\sum_{j=1}^{T'}\varphi'_j < \epsilon t\Big\}\Big)P_{\pi_D}\Big\{\sum_{j=1}^{\infty}\varphi_j I_{\{T\geq j\}} > t\Big\}.$$

Take $t_o > 0$ sufficiently large that

$$P_\nu\Big\{\sum_{j=1}^{T'}\varphi'_j < \epsilon t_o\Big\} > \frac{1}{2}.$$

In view of (A.35), Lemma A.1 applies with

$$U = \sum_{j=1}^{\infty}\varphi_j I_{\{T\geq j\}}, \quad V = \max_{n\leq m\tau_D^m}\varphi(\|S_n\|),$$

$$\beta = 1+2\epsilon, \quad \delta = 2^{-\lambda-1}\epsilon, \quad \text{and} \quad \gamma = 1 - b/2.$$

By Lemma A.3,

$$\int_D \pi(dx)E_x \max_{n\leq m\tau_D^m}\varphi(\|S_n\|) < +\infty.$$

Therefore we have proved

$$(A.38) \qquad \int_D \pi(dx)E_x\Big(\sum_{k=1}^{\infty}\varphi_k I_{\{T\geq k\}}\Big) < +\infty.$$

On the other hand, by the Markov property,

$$\int_D \pi(dx) E_x \Big(\sum_{k=1}^{\infty} \varphi_k I_{\{T \geq k\}} \Big)$$

$$= \sum_{k=1}^{\infty} \int_D \pi(dx) E_x(\varphi_k I_{\{T \geq k\}})$$

$$= \sum_{k=1}^{\infty} \int_D \pi(dx) E_x \Big(I_{\{T \geq k\}} \cdot E_{X_{m\tau_C^m(k-1)}} \varphi_1 \Big)$$

$$= \int_D \pi(dx) E_x \Big(\sum_{k=0}^{T-1} E_{X_{m\tau_C^m(k)}} \varphi_1 \Big)$$

$$= \int_C \pi(dx) E_x \varphi_1 = \int_C \pi(dx) E_x \max_{n \leq m\tau_C^m} \varphi(\|S_n\|),$$

where the fourth equality follows from Theorem 10.4.9 in Meyn-Tweedie (1993) and the fact that $\pi(C \cap \cdot)$ is the invariant measure of the Markov chain $\{X_{m\tau_C^m(k)}\}_{k \geq 0}$. Therefore, (A.31) follows from (A.38). ∎

Finally, we prove Theorem I-4.1. We need only to prove that for any $C_1, C_2 \in \mathcal{S}$, $C_1 \in \mathcal{S}_\varphi(\xi)$ implies $C_2 \in \mathcal{S}_\varphi(\xi)$. By Corollary 2.1-(iii) in Nummelin (1984) and Lemma A.4, $C_1 \cup C_2 \in \mathcal{S}_\varphi(\xi)$. Hence, by Lemma A.2 we have $C_2 \in \mathcal{S}_\varphi(\xi)$. ∎

5. Proof of Theorem I-4.2

The idea is similar to the one used in the proof of Lemma A.2. Let

(A.39) $$\tau_{\alpha^*} = \inf\{n \geq 1; \; \Phi_n \in \alpha^*\}.$$

Note that $\pi^*(\alpha^*) = b\pi(C) > 0$ and that by Proposition 5.6 in Nummelin

(1984), $0 < \pi(C) < +\infty$. We have

$$\tilde{E}_{\alpha^*} \max_{m \leq n \leq m(\tau_{\alpha^*}+1)} \varphi(\|S_n - S_m\|)$$

$$= \pi^*(\alpha^*)^{-1} \int_{\alpha^*} \pi^*(d(x,y)) \tilde{E}_{(x,y)} \max_{m \leq n \leq m(\tau_{\alpha^*}+1)} \varphi(\|S_n - S_m\|)$$

$$\leq 3^\lambda \pi^*(\alpha^*)^{-1} \Big(\int_{\alpha^*} \pi^*(d(x,y)) \tilde{E}_{(x,y)} \max_{n \leq m} \varphi(\|S_n\|)$$

$$+ \int_{\alpha^*} \pi^*(d(x,y)) \tilde{E}_{(x,y)} \max_{n \leq m\tau_{\alpha^*}} \varphi(\|S_n\|)$$

$$+ \int_{\alpha^*} \pi^*(d(x,y)) \tilde{E}_{(x,y)} \max_{m\tau_{\alpha^*} \leq n \leq m(\tau_{\alpha^*}+1)} \varphi(\|S_n - S_{m\tau_{\alpha^*}}\|) \Big)$$

$$= 3^\lambda \pi^*(\alpha^*)^{-1} \Big(2 \int_{\alpha^*} \pi^*(d(x,y)) \tilde{E}_{(x,y)} \max_{n \leq m} \varphi(\|S_n\|)$$

$$+ \int_{\alpha^*} \pi^*(d(x,y)) \tilde{E}_{(x,y)} \max_{n \leq m\tau_{\alpha^*}} \varphi(\|S_n\|) \Big),$$

where the inequality follows from the property:

$$\varphi(s_1 + s_2 + s_3) \leq 3^\lambda \big(\varphi(s_1) + \varphi(s_2) + \varphi(s_3) \big) \quad \forall s_1, s_2, s_3 \geq 0$$

and the last step from the fact that

$$\tilde{P}_{\pi_{\alpha^*}^*} \{ \Phi_{\tau_\alpha^*} \in \cdot \} = \pi_{\alpha^*}^*(\cdot), \quad (\text{where } \pi_{\alpha^*}^*(\cdot) \equiv \pi^*(\alpha^*)^{-1} \pi^*(\alpha^* \cap \cdot))$$

and the following equality resulting from the Markov property (I-2.12):

$$\int_{\alpha^*} \pi^*(d(x,y)) \tilde{E}_{(x,y)} \max_{m\tau_{\alpha^*} \leq n \leq m(\tau_{\alpha^*}+1)} \varphi(\|S_n - S_{m\tau_{\alpha^*}}\|) \Big)$$

$$= \int_{\alpha^*} \pi^*(d(x,y)) \tilde{E}_{(x,y)} \big(\tilde{E}_{\Phi_{\tau_\alpha^*}} \max_{n \leq m} \varphi(\|S_n\|) \big).$$

According to Lemma A.3, (I-4.4) implies

$$(A.40) \qquad \int_C \pi(dx) E_x \max_{n \leq m\tau_C^m} \varphi(\|S_n\|) < +\infty.$$

By definition, $m \leq m\tau_C^m$. In particular,

$$\int_{\alpha^*} \pi^*(d(x,y)) \tilde{E}_{(x,y)} \max_{n \leq m} \varphi(||S_n||)$$
$$= \pi(C) \int_{\alpha^*} \pi_C^*(d(x,y)) \tilde{E}_{(x,y)} \max_{n \leq m} \varphi(||S_n||)$$
$$\leq \pi(C) \int \pi_C^*(d(x,y)) \tilde{E}_{(x,y)} \max_{n \leq m} \varphi(||S_n||)$$
$$= \int_C \pi(dx) E_x \max_{n \leq m} \varphi(||S_n||) < +\infty,$$

where the first equality follows the fact that

(A.41) $$\pi^*((C \times I) \cap \cdot) = \pi(C) \cdot \pi_C^*(\cdot)$$

and the second from (I-2.8).

On the other hand, by (I-2.21),

$$\tilde{E}_{\nu^*} \max_{n \leq m(\tau(0)+1)} \varphi(||S_n||) = \tilde{E}_{\alpha^*} \max_{m \leq n \leq m(\tau_{\alpha^*}+1)} \varphi(||S_n - S_m||).$$

Therefore, it is enough to prove

(A.42) $$\int_{\alpha^*} \pi^*(d(x,y)) \tilde{E}_{(x,y)} \max_{n \leq m\tau_{\alpha^*}} \varphi(||S_n||) < +\infty.$$

Let the regeneration times $\{\tau_C^m(k)\}_{k \geq 0}$ be given as in (A.32) and define

$$\eta_k = \max_{m\tau_C^m(k-1) \leq n \leq m(\tau_C^m(k)+1)} ||S_n - S_{m\tau_C^m(k-1)}|| \quad k = 1, 2, \cdots$$

and

$$T = \inf\{k \geq 1; \ Y_{\tau_C^m(k)} = 1\}.$$

Then,

(A.43) $$\tau_C^m(T) = \tau_{\alpha^*} \quad \text{a.s.}$$

Choose $0 < \epsilon < 1/2$ such that

$$(A.44) \qquad 1 - \frac{b}{2(1-b)} < (1-2\epsilon)^{-\lambda}$$

and let $s \geq 0$ be fixed but arbitrary. Define

$$\rho = \inf\{k \geq 1;\ \max_{n \leq m\tau_C^m(k)} ||S_n|| > (1-2\epsilon)s\}.$$

By (A.43),

$$\tilde{P}_{\pi_{\alpha^*}^*}\{\max_{n \leq m\tau_{\alpha^*}} ||S_n|| \geq s,\ \max_{1 \leq k \leq T} \eta_k \leq \epsilon s\}$$

$$= \tilde{P}_{\pi_{\alpha^*}^*}\{\max_{n \leq m\tau_{\alpha^*}} ||S_n|| \geq s,\ \max_{1 \leq k \leq T} \eta_k \leq \epsilon s,\ \rho \leq T\}$$

$$= \sum_{l=1}^{\infty} \tilde{P}_{\pi_{\alpha^*}^*}\{\rho = l,\ T \geq l,\ \max_{n \leq m\tau_{\alpha^*}} ||S_n|| \geq s,\ \max_{1 \leq k \leq T} \eta_k \leq \epsilon s\}.$$

Let $\tilde{\theta}$ be the shift operator given in Section I-2. One can verify that for each $l \geq 1$,

$$\tau_{\alpha^*} \leq \tau_C^m(l) + \tau_{\alpha^*} \circ \tilde{\theta}^{\tau_C^m(l)} \qquad \text{a.s.}$$

and by (I-2.9) one can see

$$\max_{m(\tau_C^m(l)+1) \leq n \leq m(\tau_C^m(l)+\tau_{\alpha^*} \circ \tilde{\theta}^{\tau_C^m(l)})} ||S_n - S_{m(\tau_C^m(l)+1)}||$$

$$= \max_{m \leq n \leq m\tau_{\alpha^*}} ||S_n - S_m|| \circ \tilde{\theta}^{\tau_C^m(l)} \qquad \text{a.s.}$$

Hence,

$$\tilde{P}_{\pi_{\alpha^*}^*}\{\max_{n \leq m\tau_{\alpha^*}} ||S_n|| \geq s,\ \max_{1 \leq k \leq T} \eta_k \leq \epsilon s\}$$

$$\leq \sum_{l=1}^{\infty} \tilde{P}_{\pi_{\alpha^*}^*}\{\rho = l,\ T \geq l,\ \max_{n \leq m(\tau_C^m(l)+\tau_{\alpha^*} \circ \tilde{\theta}^{\tau_C^m(l)})} ||S_n|| \geq s,$$

$$\max_{1 \leq k \leq T} \eta_k \leq \epsilon s\}$$

$$\leq \sum_{l=1}^{\infty} \tilde{P}_{\pi_{\alpha^*}^*}\{\rho = l,\ T \geq l,$$

$$\max_{m(\tau_C^m(l)+1) \leq n \leq m(\tau_C^m(l)+\tau_{\alpha^*} \circ \tilde{\theta}^{\tau_C^m(l)})} ||S_n - S_{m(\tau_C^m(l)+1)}|| \geq \epsilon s\}$$

$$= \sum_{l=1}^{\infty} \tilde{E}_{\pi_{\alpha^*}^*}\left(I_{\{\rho=l,\ T \geq l\}} \cdot \tilde{P}_{\Phi_{\tau_C^m(l)}}\{\max_{m \leq n \leq m\tau_{\alpha^*}} ||S_n - S_m|| \geq \epsilon s\}\right).$$

Recall in the construction of the split chain, we define the kernel Q as

$$Q(x, A) = (1 - bI_C(x))^{-1}(P^m(x, A) - bI_C(x)\nu(A)) \quad x \in E, \quad A \in \mathcal{E}.$$

From (I-4.5), one can see that for all $x \in C$,

$$Q(x, \cdot) = (1 - b)^{-1}(P^m(x, \cdot) - b\nu(\cdot)) \geq b(1 - b)^{-1}\nu(\cdot).$$

Hence, $Q^*(x, \cdot) \geq b(1 - b)^{-1}\nu^*(\cdot)$ for all $x \in C$. In view of (I-2.14),

$$(A.45) \qquad \bar{P}((x, y), \cdot) \geq b(1 - b)^{-1}\nu^*(\cdot) \quad \forall (x, y) \in C \times I.$$

By the Markov property ((I-2.10)), for all $(x, y) \in C \times I$,

$$\tilde{P}_{(x,y)}\{\max_{m \leq n \leq m\tau_{\alpha^*}} \|S_n - S_m\| < \epsilon s\}$$

$$= \int \bar{P}((x, y), d(x_1, y_1))\tilde{P}_{(x,y)}\{\max_{n \leq m\tau(0)} \|S_n\| < \epsilon s\}$$

$$\geq b(1 - b)^{-1}\tilde{P}_{\nu^*}\{\max_{n \leq m\tau(0)} \|S_n\| < \epsilon s\}.$$

Consequently,

$$\tilde{P}_{\Phi_{\tau_C^m}(l)}\{\max_{m \leq n \leq m\tau_{\alpha^*}} \|S_n - S_m\| \geq \epsilon s\}$$

$$\leq 1 - b(1 - b)^{-1}\tilde{P}_{\nu^*}\{\max_{n \leq m\tau(0)} \|S_n\| < \epsilon s\} \quad \text{a.s.}$$

Therefore,

$$\tilde{P}_{\pi_{\alpha^*}}\{\max_{n \leq m\tau_{\alpha^*}} \|S_n\| \geq s, \max_{1 \leq k \leq T} \eta_k \leq \epsilon s\}$$

$$\leq \left(1 - \frac{b}{1 - b}\tilde{P}_{\nu^*}\{\max_{n \leq m\tau(0)} \|S_n\| < \epsilon s\}\right)$$

$$\times \tilde{P}_{\pi_{\alpha^*}}\{\max_{n \leq m\tau_{\alpha^*}} \|S_n\| > (1 - 2\epsilon)s\}.$$

Let $\beta = (1 - 2\epsilon)^{-\lambda}$. In view of (A.24) and (A.25), for given $t > 0$, taking $s = \varphi^{-1}(\beta t)$ we have

$$\tilde{P}_{\pi_{\alpha^*}}\{\max_{n \leq m\tau_{\alpha^*}} \varphi(\|S_n\|) \geq \beta t, \max_{1 \leq k \leq T} \varphi(\epsilon^{-1}\eta_k) < \beta t\}$$

$$\leq \left(1 - \frac{b}{1 - b}\tilde{P}_{\nu^*}\{\max_{n \leq m\tau(0)} \|S_n\| < \epsilon\varphi^{-1}(\beta t)\}\right)$$

$$\times \tilde{P}_{\pi_{\alpha^*}}\{\max_{n \leq m\tau_{\alpha^*}} \varphi(\|S_n\|) \geq t\}.$$

Take $t_o > 0$ sufficiently large such that

$$\tilde{P}_{\nu^*}\{\max_{n\leq m\tau(0)} ||S_n|| < \epsilon\varphi^{-1}(\beta t_o)\} > \frac{1}{2}.$$

In view of (A.44), Lemma A.1 applies if we take $\delta = \beta$, $\gamma = 1 - 2^{-1}b(1-b)^{-1}$, $U = \max_{n\leq m\tau_{\alpha^*}} \varphi(||S_n||)$ and $V = \max_{1\leq k\leq T} \varphi(\epsilon^{-1}\eta_k)$. Therefore, it remains to prove

$$\int_{\alpha^*} \pi^*(d(x,y))\tilde{E}_{(x,y)} \max_{1\leq k\leq T} \varphi(\epsilon^{-1}\eta_k) < +\infty.$$

By (A.14), we need only to prove

$$(A.46) \qquad \int_{\alpha^*} \pi^*(d(x,y))\tilde{E}_{(x,y)} \sum_{k=1}^{T} \varphi(\eta_k) < +\infty.$$

In fact,

$$\int_{\alpha^*} \pi^*(d(x,y))\tilde{E}_{(x,y)} \sum_{k=1}^{T} \varphi(\eta_k)$$

$$= \int_{\alpha^*} \pi^*(d(x,y))\tilde{E}_{(x,y)} \sum_{k=0}^{T-1} \tilde{E}_{\Phi_{\tau_C^m(k)}}\big(\max_{n\leq m(\tau_C^m+1)} \varphi(||S_n||)\big)$$

$$= \int_{C\times I} \pi^*(d(x,y))\tilde{E}_{(x,y)}\big(\max_{n\leq m(\tau_C^m+1)} \varphi(||S_n||)\big)$$

$$= \pi(C)\int \pi_C^*(d(x,y))\tilde{E}_{(x,y)}\big(\max_{n\leq m(\tau_C^m+1)} \varphi(||S_n||)\big)$$

$$= \pi(C)\int \pi_C(dx)E_x\big(\max_{n\leq m(\tau_C^m+1)} \varphi(||S_n||)\big)$$

$$= \int_C \pi(dx)E_x\big(\max_{n\leq m(\tau_C^m+1)} \varphi(||S_n||)\big),$$

where the first step follows from the Markov property, the second from Theorem 10.4.9 in Meyn-Tweedie (1993) and the fact that $\pi^*((C\times I)\cap \cdot)$ is the invariant measure of the Markov chain $\{\Phi_{\tau_C^m(k)}\}_{k\geq 0}$, the third from (A.41) and the fourth from (I-2.8). From the argument used in the proof of (A.27), one can see how (A.40) implies

$$\int_C \pi(dx) E_x\big(\max_{n \leq m(\tau_C^m + 1)} \varphi(||S_n||) \big) < +\infty.$$

Hence, we have (A.46). ∎

REFERENCES

[1]. de Acosta, A. (1981). Inequalities for B-valued random vectors with applications to the strong law of large numbers. Ann. Probab. **9** 157-161.

[2]. de Acosta, A. (1983). Small deviations in the functional central limit theorem with applications to functional laws of the iterated logarithm. Ann. Probab. **11** 78-101.

[3]. de Acosta, A. (1988). Large deviations for vector-valued functional of a Markov chain: lower bounds. Ann. Probab. **16** 925-960.

[4]. de Acosta, A. (1997). Moderate deviations for empirical measures of Markov chains: lower bounds. Ann. Probab. **25** 259-284.

[5]. de Acosta, A. and Chen, X. (199?). Moderate deviations for empirical measures of Markov chains: upper bounds. J. Theor. Probab. (to appear).

[6]. de Acosta, A. and Kuelbs, J. (1983). Limit theorems for moving averages of independent random vectors. Z. Wahrs. verw Gebiete **64** 67-123.

[7]. de Acosta, A. and Kuelbs, J. (1983). Some results on the cluster set $C(\{S_n/a_n\})$ and the LIL. Ann. Probab. **11** 102-122.

[8]. Athreya, K. B. and Ney, P. (1978). A new approach to the limit theory of recurrent Markov chains. Trans. Amer. Math. Soc. **245** 493-501.

[9]. Billingsley, P. (1968). Convergence of Probability Measures. Wiley. New York.

[10]. Burkholder, D. L. and Gundy, R. F. (1970). Extrapolation and interpolation of quasi-linear operators on martingales. Acta Math. **124** 249-304.

[11]. Chen, X. (1990). Probabilities of moderate deviations for B-valued independent random vectors. Chinese Journal of Contemporary Mathematics. Vol. **11** 381-393, Allerton Press Inc. New York.

[12]. Chen, X. (1991). Probabilities of moderate deviations for independent random vectors in a Banach space. Chinese J. of Appl.

Probab. and Statis. **7** 24-32.

[13]. Chen, X. (1993). On the law of the iterated logarithm for independent Banach space valued random variables. Ann Probab. **21** 1991-2011.

[14]. Chen, X.(1997).The law of the iterated logarithm for m-dependent Banach space valued random variables. J. Theor. Probab. **10** 695-732.

[15]. Chen, X. (1997). Moderate deviations for m-dependent random variables with Banach space values. Stat.&Probab. Letters. **35** 123-134.

[16]. Chow, Y. S. and Teicher, H. (1988). Probability Theory, Independence, Interchangeability, Martingales. Springer-Verlag, New York.

[17]. Chung, K. L. (1967). Markov Chains with Stationary Transition Probabilities. Springer-Verlag, Berlin, 2nd edition.

[18]. Cogburn, R. (1972). The central limit theorem for Markov processes. In Procedings of the 6th Berkeley Symposium on Mathematical Statistics and Probability, University of California Press, 485-512.

[19]. Csáki, E. and Csörgő, M. (1995). On additive functionals of Markov chains. J. Theor. Probab. **8** 905-919.

[20]. Dembo, A. and Zeitouni, O. (1993). Large Deviations Techniques and Applications. Jones and Bartlett, Boston and London.

[21]. Deuschel, J. D. and Stroock, D. W. (1989). Large Deviations. Academic Press, Boston.

[22]. Dinwoodie, I. H. and Ney, P. (1995). Occupation measures for Markov chains. J. Theor. Probab. **8** 679-691.

[23]. Dunford, N. and Schwartz, J. T. (1964). Linear Operators (I). Interscience, New York.

[24]. Gao, F. Q.(1994).Uniform moderate deviations for Markov processes. Research Announcements, Advances in Mathematics (China).

[25]. Giné, E. and Zinn, J. (1984). Some limit theorems for empirical

processes. Ann. Probab. **12** 929-989.

[26]. Goodman, V., Kuelbs, J. and Zinn, J. (1981). Some results on the LIL in Banach space with applications to weighted empirical processes. Ann. Probab. **9** 713-752.

[27]. Hall, P. and Heyde, C. C. (1980). Martingale Limit Theorems and Its application. Academic Press, New York.

[28]. Klass, M. J. (1988). A best possible improvement of Wald's equation. Ann. Probab. **16** 840-653.

[29]. Klass, M. J. (1990). Uniform lower bounds for randomly stopped Banach space valued random sums. Ann. Probab. **18** 790-809.

[30]. Kuelbs, J. (1981). When is the cluster set of $\{\frac{S_n}{a_n}\}$ empty? Ann. Probab. **9** 377-394.

[31]. Ledoux, M. and Talagrand, M. (1991). Probability in Banach Spaces. Springer, Berlin.

[32]. Maigret, N. (1978). Théorème de limite centrale pour une chaine de Markov récurrente Harris positive. Ann. Inst. Henri Poincaré. Ser B **14** 425-440.

[33]. Meyn, S. P. and Tweedie, R. L. (1993). Markov Chains and Stochstic Stability. Springer-Verlag. London.

[34]. Mogulskii, A. A. (1984). On moderately large deviations from the invariant measure. Advances in Probability Theory: limit theorems and related problems. A. A. Borovkov ed.. Optimization Software, New York.

[35]. Ney, P. and Nummelin, E. (1987). Markov additive processes I. Eigenvalue properties and limit theorems. Ann. Probab. **15** 561-592.

[36]. Ney, P. and Nummelin, E. (1987). Markov additive processes II. Large deviations. Ann. Probab. **15** 593-609.

[37]. Niemi, S. and Nummelin, E. (1982). Central limit theorems for Markov random walks. Commentationes Physico-Mathematicae. **54**, Societas Scientiarum Fennica, Helsinki.

[38]. Nummelin, E. (1978). A splitting technique for Harris recurrent

chains. Z. Wahrs. verw Gebiete **43** 309-318.

[39]. Nummelin, E. (1984). General Irreducible Markov Chains and Non-negative Operators. Cambridge University Press, Cambridge, England.

[40]. Revuz, D. (1975). Markov Chains. North-Holland, New York.

[41]. Stout, W. F. (1970). The Hartman-Wintner law of the iterated logarithm for martingales. Ann. Math. Statist. **41** 2158-2160.

[42]. Stroock, D. W. (1984). An Introduction to the Theory of Large Deviations. Springer-Verlag, Berlin.

[43]. Touati, A. (1990). Loi functionelle du logarithme itéré pour les processus de Markov récurrents. Ann. Probab. **18** 140-159.

[44]. Wu, L. M. (1995). Moderate deviations of dependent random variables related to CLT. Ann. Probab. **23** 420-445.

Editorial Information

To be published in the *Memoirs*, a paper must be correct, new, nontrivial, and significant. Further, it must be well written and of interest to a substantial number of mathematicians. Piecemeal results, such as an inconclusive step toward an unproved major theorem or a minor variation on a known result, are in general not acceptable for publication. *Transactions* Editors shall solicit and encourage publication of worthy papers. Papers appearing in *Memoirs* are generally longer than those appearing in *Transactions* with which it shares an editorial committee.

As of January 31, 1999, the backlog for this journal was approximately 4 volumes. This estimate is the result of dividing the number of manuscripts for this journal in the Providence office that have not yet gone to the printer on the above date by the average number of monographs per volume over the previous twelve months, reduced by the number of issues published in four months (the time necessary for preparing an issue for the printer). (There are 6 volumes per year, each containing at least 4 numbers.)

A Copyright Transfer Agreement is required before a paper will be published in this journal. By submitting a paper to this journal, authors certify that the manuscript has not been submitted to nor is it under consideration for publication by another journal, conference proceedings, or similar publication.

Information for Authors and Editors

Memoirs are printed by photo-offset from camera copy fully prepared by the author. This means that the finished book will look exactly like the copy submitted.

The paper must contain a *descriptive title* and an *abstract* that summarizes the article in language suitable for workers in the general field (algebra, analysis, etc.). The *descriptive title* should be short, but informative; useless or vague phrases such as "some remarks about" or "concerning" should be avoided. The *abstract* should be at least one complete sentence, and at most 300 words. Included with the footnotes to the paper, there should be the 1991 *Mathematics Subject Classification* representing the primary and secondary subjects of the article. This may be followed by a list of *key words and phrases* describing the subject matter of the article and taken from it. A list of the numbers may be found in the annual index of *Mathematical Reviews*, published with the December issue starting in 1990, as well as from the electronic service e-MATH [**telnet e-MATH.ams.org** (or **telnet 130.44.1.100**). Login and password are **e-math**]. For journal abbreviations used in bibliographies, see the list of serials in the latest *Mathematical Reviews* annual index. When the manuscript is submitted, authors should supply the editor with electronic addresses if available. These will be printed after the postal address at the end of each article.

Electronically prepared papers. The AMS encourages submission of electronically prepared papers in $\mathcal{A}_{\mathcal{M}}\mathcal{S}$-TeX or $\mathcal{A}_{\mathcal{M}}\mathcal{S}$-LaTeX. The Society has prepared author packages for each AMS publication. Author packages include instructions for preparing electronic papers, the *AMS Author Handbook*, samples, and a style file that generates the particular design specifications of that publication series for both $\mathcal{A}_{\mathcal{M}}\mathcal{S}$-TeX and $\mathcal{A}_{\mathcal{M}}\mathcal{S}$-LaTeX.

Authors with FTP access may retrieve an author package from the Society's Internet node `e-MATH.ams.org` (130.44.1.100). For those without FTP access, the author package can be obtained free of charge by sending e-mail to **pub@ams.org** (Internet) or from the Publication Division, American Mathematical Society, P.O. Box 6248, Providence, RI 02940-6248. When requesting an author package, please specify $\mathcal{AMS}$-TeX or $\mathcal{AMS}$-LaTeX, Macintosh or IBM (3.5) format, and the publication in which your paper will appear. Please be sure to include your complete mailing address.

Submission of electronic files. At the time of submission, the source file(s) should be sent to the Providence office (this includes any TeX source file, any graphics files, and the DVI or PostScript file).

Before sending the source file, be sure you have proofread your paper carefully. The files you send must be the EXACT files used to generate the proof copy that was accepted for publication. For all publications, authors are required to send a printed copy of their paper, which exactly matches the copy approved for publication, along with any graphics that will appear in the paper.

TeX files may be submitted by email, FTP, or on diskette. The DVI file(s) and PostScript files should be submitted only by FTP or on diskette unless they are encoded properly to submit through e-mail. (DVI files are binary and PostScript files tend to be very large.)

Files sent by electronic mail should be addressed to the Internet address **pub-submit@ams.org**. The subject line of the message should include the publication code to identify it as a Memoir. TeX source files, DVI files, and PostScript files can be transferred over the Internet by FTP to the Internet node **e-math.ams.org** (130.44.1.100).

Electronic graphics. Figures may be submitted to the AMS in an electronic format. The AMS recommends that graphics created electronically be saved in Encapsulated PostScript (EPS) format. This includes graphics originated via a graphics application as well as scanned photographs or other computer-generated images.

If the graphics package used does not support EPS output, the graphics file should be saved in one of the standard graphics formats—such as TIFF, PICT, GIF, etc.—rather than in an application-dependent format. Graphics files submitted in an application-dependent format are not likely to be used. No matter what method was used to produce the graphic, it is necessary to provide a paper copy to the AMS.

Authors using graphics packages for the creation of electronic art should also avoid the use of any lines thinner than 0.5 points in width. Many graphics packages allow the user to specify a "hairline" for a very thin line. Hairlines often look acceptable when proofed on a typical laser printer. However, when produced on a high-resolution laser imagesetter, hairlines become nearly invisible and will be lost entirely in the final printing process.

Screens should be set to values between 15% and 85%. Screens which fall outside of this range are too light or too dark to print correctly.

Any inquiries concerning a paper that has been accepted for publication should be sent directly to the Editorial Department, American Mathematical Society, P. O. Box 6248, Providence, RI 02940-6248.

Editors

This journal is designed particularly for long research papers (and groups of cognate papers) in pure and applied mathematics. Papers intended for publication in the *Memoirs* should be addressed to one of the following editors:

Ordinary differential equations, partial differential equations, and applied mathematics to JOHN MALLET-PARET, Division of Applied Mathematics, Brown University, Providence, RI 02912-9000; electronic mail: `jmp@cfm.brown.edu`.

Harmonic analysis, representation theory, and Lie theory to ROBERT J. STANTON, Department of Mathematics, The Ohio State University, 231 West 18th Avenue, Columbus, OH 43210-1174; electronic mail: `stanton@math.ohio-state.edu`.

Ergodic theory and dynamical systems to ROBERT F. WILLIAMS, Department of Mathematics, University of Texas at Austin, Austin, TX 78712-1082; e-mail: `bob@math.utexas.edu`

Real and harmonic analysis and geometric partial differential equations to WILLIAM BECKNER, Department of Mathematics, University of Texas at Austin, Austin, TX 78712-1082; e-mail: `beckner@math.utexas.edu`.

Algebra to CHARLES CURTIS, Department of Mathematics, University of Oregon, Eugene, OR 97403-1222 e-mail: `cwc@darkwing.uoregon.edu`

Algebraic topology and cohomology of groups to STEWART PRIDDY, Department of Mathematics, Northwestern University, 2033 Sheridan Road, Evanston, IL 60208-2730; e-mail: `s_priddy@math.nwu.edu`.

Differential geometry and global analysis to CHUU-LIAN TERNG, Department of Mathematics, Northeastern University, Huntington Avenue, Boston, MA 02115-5096; e-mail: `terng@neu.edu`.

Probability and statistics to RODRIGO BAÑUELOS, Department of Mathematics, Purdue University, West Lafayette, IN 47907-1968; e-mail: `banuelos@math.purdue.edu`.

Combinatorics and Lie theory to PHILIP J. HANLON, Department of Mathematics, University of Michigan, Ann Arbor, MI 48109-1003; e-mail: `hanlon@math.lsa.umich.edu`.

Logic to THEODORE SLAMAN, Department of Mathematics, University of California at Berkeley, Berkeley, CA 94720-3840; e-mail: `slaman@math.berkeley.edu`.

Number theory and arithmetic algebraic geometry to ALICE SILVERBERG, MSRI, 1000 Centennial Dr., Berkeley, CA 94720; e-mail: `silver@math.ohio-state.edu`.

Complex analysis and complex geometry to DANIEL M. BURNS, Department of Mathematics, University of Michigan, Ann Arbor, MI 48109-1003; e-mail: `dburns@math.lsa.umich.edu`.

Algebraic geometry and commutative algebra to LAWRENCE EIN, Department of Mathematics, University of Illinois, 851 S. Morgan (M/C 249), Chicago, IL 60607-7045; e-mail: `ein@uic.edu`.

Geometric topology, knot theory, hyperbolic geometry, and general topoogy to JOHN LUECKE, Department of Mathematics, University of Texas at Austin, Austin, TX 78712-1082; e-mail: `luecke@math.utexas.edu`.

Partial differential equations and applied mathematics to BARBARA LEE KEYFITZ, Department of Mathematics, University of Houston, 4800 Calhoun, Houston, TX 77204-3476; e-mail: `keyfitz@uh.edu`

Operator algebras and functional analysis to BRUCE E. BLACKADAR, Department of Mathematics, University of Nevada, Reno, NV 89557; e-mail: `bruceb@math.unr.edu`

All other communications to the editors should be addressed to the Managing Editor, PETER SHALEN, Department of Mathematics, University of Illinois, 851 S. Morgan (M/C 249), Chicago, IL 60607-7045; e-mail: `shalen@math.uic.edu`.

Selected Titles in This Series

(Continued from the front of this publication)

635 **Magdy Assem,** On stability and endoscopic transfer of unipotent orbital integrals on p-adic symplectic groups, 1998

634 **Darrin D. Frey,** Conjugacy of Alt_5 and $SL(2,5)$ subgroups of $E_8(\mathbb{C})$, 1998

633 **Dikran Dikranjan and Dmitri Shakhmatov,** Algebraic structure of pseudocompact groups, 1998

632 **Shouchuan Hu and Nikolaos S. Papageorgiou,** Time-dependent subdifferential evolution inclusions and optimal control, 1998

631 **Ronnie Lee, Steven H. Weintraub, and J. William Hoffman,** The Siegel modular variety of degree two and level four/Cohomology of the Siegel modular group of degree two and level four, 1998

630 **Florin Rădulescu,** The Γ-equivariant form of the Berezin quantization of the upper half plane, 1998

629 **Richard B. Sowers,** Short-time geometry of random heat kernels, 1998

628 **Christopher K. McCord, Kenneth R. Meyer, and Quidong Wang,** The integral manifolds of the three body problem, 1998

627 **Roland Speicher,** Combinatorial theory of the free product with amalgamation and operator-valued free probability theory, 1998

626 **Mikhail Borovoi,** Abelian Galois cohomology of reductive groups, 1998

625 **George Xian-Zhi Yuan,** The study of minimax inequalities and applications to economies and variational inequalities, 1998

624 **P. Deift and K. T-R McLaughlin,** A continuum limit of the Toda lattice, 1998

623 **S. A. Adeleke and Peter M. Neumann,** Relations related to betweenness: Their structure and automorphisms, 1998

622 **Luigi Fontana, Steven G. Krantz, and Marco M. Peloso,** Hodge theory in the Sobolev topology for the de Rham complex, 1998

621 **Gregory L. Cherlin,** The classification of countable homogeneous directed graphs and countable homogeneous n-tournaments, 1998

620 **Victor Guba and Mark Sapir,** Diagram groups, 1997

619 **Kazuyoshi Kiyohara,** Two classes of Riemannian manifolds whose geodesic flows are integrable, 1997

618 **Karl H. Hofmann and Wolfgang A. F. Ruppert,** Lie groups and subsemigroups with surjective exponential function, 1997

617 **Robin Hartshorne,** Families of curves in $\mathbb{P}^3$ and Zeuthen's problem, 1997

616 **Serguei G. Bobkov and Christian Houdré,** Some connections between isoperimetric and Sobolev-type inequalities, 1997

615 **Michael A. Dritschel and Hugo J. Woerdeman,** Model theory and linear extreme points in the numerical radius unit ball, 1997

614 **Richard Warren,** The structure of k-CS-transitive cycle-free partial orders, 1997

613 **D. L. Flannery,** The finite irreducible linear 2-groups of degree 4, 1997

612 **Joan Porti,** Torsion de Reidemeister pour les variétés hyperboliques, 1997

611 **D. Ginzburg, I. Piatetski-Shapiro, and S. Rallis,** L functions for the orthogonal group, 1997

610 **Mark Hovey, John H. Palmieri, and Neil P. Strickland,** Axiomatic stable homotopy theory, 1997

609 **Liviu I. Nicolaescu,** Generalized symplectic geometries and the index of families of elliptic problems, 1997

608 **Christina Q. He and Michel L. Lapidus,** Generalized Minkowski content, spectrum of fractal drums, fractal strings, and the Riemann zeta-functions, 1997

607 **Adele Zucchi,** Operators of class C_0 with spectra in multiply connected regions, 1997

(See the AMS catalog for earlier titles)